Oil and Gas Engineering for Non-Engineers

Oil and Gas Engineering for Non-Engineers explains in non-technical terms how oil and gas exploration and production are carried out in the upstream oil and gas industry. The aim is to help readers with no prior knowledge of the oil and gas industry obtain a working understanding of the field.

- Focuses on just the basics of what the layperson needs to know to understand the industry
- Uses non-technical terms, simple explanations, and illustrations to describe the inner workings of the field
- Explains how oil is detected underground, how well locations are determined, how drilling is done, and how wells are monitored during production
- Describes how and why oil and gas are separated from impurities before being sent to customers

Aimed at non-engineers working within the oil and gas sector, this book helps readers to get comfortable with the workings of this advanced field without the need for an advanced degree in the subject.

Oil and Gas Engineering for Non-Engineers

Quinta Nwanosike-Warren

CRC Press
Taylor & Francis Group
Boca Raton London New York

CRC Press is an imprint of the
Taylor & Francis Group, an **informa** business

First edition published 2023
by CRC Press
6000 Broken Sound Parkway NW, Suite 300, Boca Raton, FL 33487-2742

and by CRC Press
2 Park Square, Milton Park, Abingdon, Oxon, OX14 4RN

ISBN: 978-0-367-60772-2 (hbk)
ISBN: 978-0-367-60769-2 (pbk)
ISBN: 978-1-003-10046-1 (ebk)

DOI: 10.1201/9781003100461

Typeset in Palatino
by MPS Limited, Dehradun

Contents

List of Figures

List of Tables

Preface

This textbook was written to provide insight into the engineering processes employed to extract oil and gas. It is useful for anyone who has ever wondered how oil and gas are discovered, and how they make their way into our gas tanks and barbecue grills. Processes are explained in layman's terms so non-engineers can easily understand them. Terms are explained when they are used.

It should be noted that there are many disciplines involved in oil and gas extraction. Although this textbook focuses on the engineering disciplines involved in getting oil and gas out of the ground, the foundation of the process lies in geology, so one chapter is dedicated to the basics of geology and how they relate to the oil and gas industry. In particular, reservoir, drilling, and completions engineers need a basic understanding of geology to be effective in their roles.

More emphasis is placed on the extraction process or upstream oil and gas. However, transportation and refining of crude oil are also lightly addressed to provide a more complete picture of the oil and gas industry.

Author

Dr. Quinta Nwanosike-Warren, PE, PMP, is a chemical engineer and an energy professional. She is the Founder and CEO of Energy Research Consulting (EngrRC.com), which helps accelerate entrepreneurial ventures in Africa by providing energy expertise. The company focuses on solutions that are sustainable, fit-for-purpose, and tailored for cultural and local context.

Dr. Nwanosike-Warren has worked for Consumer Reports leading sustainability policy work, and for ConocoPhillips on CO_2 capture research, and as a reservoir engineer for heavy oil and tight gas assets. She was an American Association for the Advancement of Science (AAAS) Fellow with the US Department of Energy, contributing to domestic and international policy around carbon management and power generation/ transmission/distribution. She was also an AAAS Fellow with the Millennium Challenge Corporation (MCC), a US government agency that works on international development. At MCC, she provided technical energy expertise and contributed to the development of electricity projects in emerging economies in Asia and Africa.

Dr. Nwanosike-Warren holds a PhD in Chemical & Biomolecular Engineering from Georgia Tech and a Bachelor's degree in Chemical Engineering from Penn State University.

Abbreviations

Acronym	Meaning
Bbl	Barrel
BOE	Barrels of Oil Equivalent
BOPD	Barrels of Oil Per Day
CBM	Coal Bed Methane
CO_2	Carbon Dioxide
EOR	Enhanced Oil Recovery
EUR	Estimated Ultimate Recovery
FCC	Fluid Catalytic Cracker
G&A	General and Administrative
IRR	Internal Rate of Return
LiDAR	Light Detection and Ranging
LNG	Liquified Natural Gas
LOE	Lease Operating Expenses
LPG	Liquified Petroleum Gas
LWD	Logging While Drilling
MMBOPD	Million Barrels of Oil Per Day
MMCF	Million Cubic Feet
MWD	Measurement While Drilling
NGL	Natural Gas Liquids
NPV	Net Present Value
OGIP	Original Gas in Place
OHIP	Original Hydrocarbons in Place
OOIP	Original Oil in Place
OPEC	Organization of Petroleum Exporting Countries
P&A	Plugging and Abandonment
RF	Recovery Factor
SAGD	Steam-Assisted Gravity Drainage
SEC	Securities and Exchange Commission
SPR	Strategic Petroleum Reserves
STB	Stock Tank Barrel
TLP	Tension Leg Platform
WTI	West Texas Intermediate

1

Introduction

1.1 What Is Crude Oil?

Oil is an integral part of our world today. It is in the clothes we wear, our cellphones, cosmetics, medication, and more. Let us not forget how important it is for transportation whether by land, air, or sea. It is the primary source of energy in the world, and as such, it is the most important commodity in the world.

Gas is usually produced alongside crude oil as a by-product. The composition of this associated gas can vary but often includes methane, also called natural gas, and propane. In some cases, associated gas is sold to help with the production of the crude oil. In other cases, it is an undesirable product so it is vented into the air, flared or burned in a controllable way, or re-injected into the reservoir. Flaring is increasingly being viewed as unsustainable and, of course, bad for the environment due to the global warming properties of methane. In some countries such as Nigeria, gas flaring is no longer allowed, leading to development of gas utilization projects. Natural gas can also be produced on its own from natural gas reservoirs. In this case, the gas is called non-associated gas.

Oil and gas are used to produce materials, fuels, electricity/heat, and petrochemicals. Materials include plastics and synthetic fabrics such as polyester, nylon, and rayon. Petrochemicals include fertilizers, medication, paint, and cosmetics. Fuels such as gasoline, diesel, and jet fuel are primarily used for transportation for cars, planes, and ships. In electricity generation, fuels such as natural gas, and heavy fuel oil, are combusted to generate steam. Uses of oil and gas are illustrated in Figure 1.1.

Crude oil is refined into different products by heating. At different temperatures, different fuels boiloff from the crude oil mixture. The typical products of refining are fuels, naphtha, and residue. The fuels include propane, butane, gasoline, jet fuel, kerosene, diesel, and fuel oil. Naphtha is used as a precursor for other chemicals in the petrochemical industry. The residual materials are asphalt, and lube stock which is used for lubricants, waxes, and polishes. Products of crude oil refining are illustrated in Figure 1.2.

DOI: 10.1201/9781003100461-1

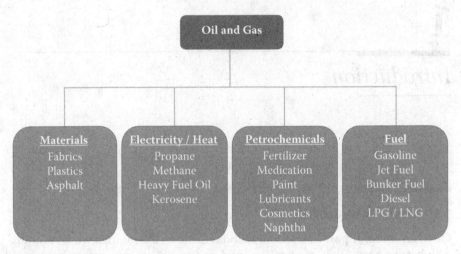

FIGURE 1.1
Products from oil and gas.

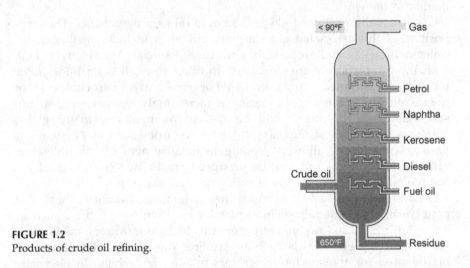

FIGURE 1.2
Products of crude oil refining.

1.2 How Oil Is Formed

Petroleum, or crude oil, is a liquid mixture of hydrocarbons formed underground from the remains of plants and marine life such as algae and plankton. It is a mixture of compounds called hydrocarbons which are made up of mostly carbon and hydrogen. The color of crude oil can vary from yellow to black depending on its composition: heavier hydrocarbons will lead to a darker colored crude.

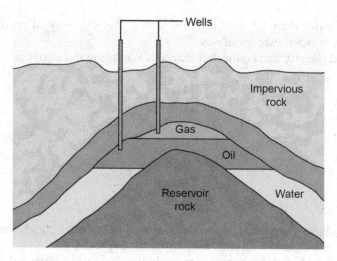

FIGURE 1.3
Oil and gas production from hydrocarbon reservoirs underground.

The plankton remains are buried and subjected to high pressures and temperatures due to the depth of burial. The deeper the remains are buried, the higher the temperature and pressure. The longer the remains are subjected to these conditions, the more likely they will be converted to natural gas rather than crude oil. The hydrocarbons are formed in source rock and then migrate into shallower rock called reservoir rock. It is from reservoir rock that most crude oil is produced. The exception is shale oil and gas which are produced from source rock.

Wells are drilled into the reservoir as shown in Figure 1.3 to create a conduit through which oil and gas can be produced. In the reservoir, gas is usually found at the top because it is light, followed by oil in the middle and water at the bottom because it is heaviest. Reservoir depths may vary from as shallow as 3,500 ft to as deep as 12,000 ft.

1.3 Classification of Crude Oil

Crude oil may be classified in a number of ways either based on the physical characteristics of the crude or based on the reservoir it is extracted from. One classification method is on the density of the oil. API gravity is a measure of how dense a hydrocarbon is compared to water, and it is measured in degrees. The smaller the API of a petroleum liquid, the denser it is, with extra heavy crude oil being the densest.

Crude oil with an API of 10° has a density the same as that of water. Crude oil with an API less than 10° will sink in water while crude oil with an API greater than 10° will float on water. The viscosity of crude oil increases as temperature increases.

API and density are connected through the following equation.

$$API = \frac{141.5}{Specific\ Gravity} - 131.5 \qquad\qquad 1.1$$

where

$$Specific\ Gravity = \frac{\rho_{liquid}}{\rho_{water}} \qquad\qquad 1.2$$

and ρ_{liquid} is the density of the crude oil liquid in kg/m^3, ρ_{water} is the density of water, which is 999 kg/m^3, and API is in degrees (°).

The API and density ranges for each type of crude vary for different countries and different sources. One set of ranges is given in Table 1.1.

Examples of extra heavy crude oil include bitumen and tar sands or oil sands. Light crude such as shale oil is easy to refine and is therefore more desirable and more expensive.

Another way of classifying oil is by sulfur content. The oil that contains less than 1% sulfur is called sweet crude. Sour crude is crude that contains greater than 1% sulfur. Sweet crude is more valuable because sulfur is undesirable and must be removed from sour crude. West Texas Intermediate (WTI) is a light sweet crude sourced from inland Texas' Permian Basin. It is used as a benchmark or reference for crude oil pricing in North America. Brent crude is a light sweet crude from the North Sea between the United Kingdom and Norway which is used for benchmarking two-thirds of crude oil traded worldwide. It is less light and less sweet than WTI but is more widely used for benchmarking because it is the most traded crude in the world.

The North American natural gas benchmark is Henry Hub. It is named after a pipeline interchange located in Louisiana which moves gas from across the Gulf Coast. Other natural gas benchmarks include National Balancing Point (NBP) in the United Kingdom and Title Transfer Facility (TTF) in the Netherlands.

TABLE 1.1

Oil types and their associated APIs and densities at reservoir conditions

Oil type	API (°)	Density (kg/m^3)
Extra heavy crude oil	<10	>999
Heavy oil	10–20	933–1000
Conventional oil	20–40	824–933
Light crude oil	>40	<824

Yet another way of classifying petroleum is by conventional and unconventional reservoirs. Unconventional reservoirs can refer to any way of recovering oil and gas that is outside of the norm. Thus, shale oil and gas, tight gas, heavy oil, and deepwater reservoirs all count as unconventional. Tight gas reservoirs are characterized by low porosity and low permeability, requiring fracking in order for natural gas to be produced. Porosity is the percentage of void space in a rock that can hold liquids or gases. The higher the porosity, the greater the ability of the rock to hold crude oil and other fluids. Shale oil and gas are similarly found in low permeability rock, but instead of the oil and gas being found in reservoir rock, it is found in source rock.

Heavy oil is solid at room temperature and so it cannot be produced like conventional oil as it does not flow under normal conditions. Steam has to be pumped into the reservoir to reduce the viscosity of heavy oil so it can flow and can be produced. If the reservoir is shallow, say around 500 ft, the heavy oil may be produced using surface mining methods. Deepwater oil requires special oil rigs to be used which can withstand the heavy waves and sandy floors of the ocean. Unconventional reservoirs are typically more expensive to produce from than conventional reservoirs.

1.4 Petroleum Value Chain

1.4.1 History of Petroleum Use

Petroleum has been in use since time immemorial. Asphalt is recorded as being in use for the construction of walls in Babylon as far back as 4,000 years ago. The first recorded oil wells were drilled around 347 AD in China at depths of around 800 ft. Refining of crude oil was being carried out in the 9th century by Persian scientists.

In other parts of the world, whale blubber was found to be particularly useful for lighting. With the invention of the internal combustion engine in the late 1800s, demand for oil skyrocketed leading to modern-day oil production.

In 1859, the first commercial oil well in the United States was drilled in Titusville, Pennsylvania. The success of this well led to more widespread production of oil, and its replacement of other oil sources such as whale blubber. Wildcatting became common, i.e., using wild guesses to determine where to drill for oil. Over time, principles tied to existence of oil were established and the disciplines of geology and reservoir engineering were born. Today, drilling for hydrocarbons is a sophisticated science using sophisticated tools and instruments, and involving multiple disciplines that must work together to achieve the common goal of oil and gas production.

1.4.2 Impact of Oil on World Economies

The world economy is closely tied to the availability and price of crude oil because, as earlier mentioned, oil is used in almost every aspect of our lives. Disruption of production in one country can affect prices throughout the world. For instance, the Arab Spring in North Africa and the Middle East in the early 2010s saw oil supply disrupted in countries such as Egypt, Libya, and led to prices increasing worldwide. World consumption change, world GDP change, and WTI crude oil prices from 2002 to 2021 are shown in Figure 1.4. Oil consumption fell sharply in 2020 due to the COVID-19 pandemic, and this caused world GDP and oil prices to fall sharply as well.

High oil prices are beneficial for oil-producing countries as their budgets tend to depend heavily on exporting oil. However, high oil prices lead to increased transportation costs and increased raw material costs for some industries. This in turn leads to higher prices of products in almost every sector.

As shown in Table 1.1, the top 10 oil-producing countries in the world in 2020 were the United States, Saudi Arabia, Russia, Canada, China, Iraq, the UAE, Brazil, Iran, and Kuwait. Saudi Arabia was historically the world's largest producer but was overtaken by the United States in 2015 due to the shale oil revolution. The top 10 oil producers in 2020 accounted for 72% of total world oil production, which was 93.86 million barrels of oil/day (Table 1.2).

In 2020, the Middle East was the biggest producer of oil with 27.7 million barrels of oil per day (MMBOPD), and North America was a close second

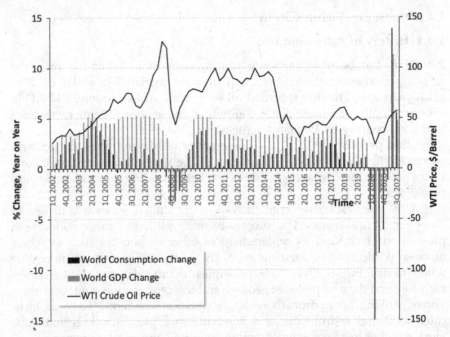

FIGURE 1.4
WTI crude oil price, world GDP growth, and oil consumption growth over time[1].

TABLE 1.2

Top 10 oil-producing countries in 2020[2]

Country	Million barrels per day	Share of the world total (%)
United States	18.61	20
Saudi Arabia	10.81	12
Russia	10.50	11
Canada	5.23	6
China	4.86	5
Iraq	4.16	4
United Arab Emirates	3.78	4
Brazil	3.77	4
Iran	3.01	3
Kuwait	2.75	3
TOTAL	67.49	72

with 23.5 MMBOPD. The biggest consumer of oil in 2020 was Asia Pacific with 33.6 MMBOPD, while North America was second with 20.8 MMBOPD. North America has historically led in oil consumption due to high oil consumption in the United States. However, US oil consumption has been level or decreasing in the last decade while China's oil consumption has been steadily decreasing. Figure 1.5 shows world oil production and consumption by different regions of the world from 1965 to 2020.

The Organization of Petroleum Exporting Countries (OPEC) is a multi-lateral organization composed of 13-member oil-exporting countries. It was founded to minimize fluctuations in international oil prices. OPEC members' share of world crude oil reserves in 2018 was 79% as shown in Figure 1.6, with Venezuela in first place with 302 billion barrels of crude oil. The member countries of OPEC are Algeria, Angola, the Republic of Congo, Equatorial Guinea, Gabon, Iran, Iraq, Kuwait, Libya, Nigeria, Saudi Arabia, United Arab Emirates, and Venezuela. Ecuador and Qatar withdrew from the organization in 2020 and 2019, respectively. Indonesia suspended its membership in 2016.

Venezuela has seen production decline since 2015 from about 2,500,000 barrels of oil per day (BOPD) to about 360,000 BOPD in 2020 due to mismanagement of oil production in the country as well as sanctions put on the country by the United States, leading to Venezuela not being able to sell its oil internationally. OPEC as a whole works together to prevent flooding the market with too much oil. Saudi Arabia is the biggest oil producer in OPEC. As a result, agreements to reduce production largely depend on Saudi Arabia's participation.

Some countries keep an emergency supply of crude oil to minimize disruptions in oil supply. In the United States, this emergency supply is called the Strategic Petroleum Reserves (SPR). It is stored underground in salt

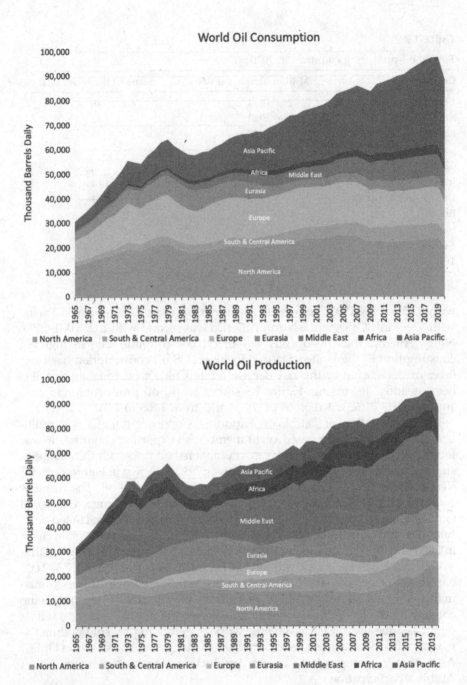

FIGURE 1.5
Oil production and consumption by world regions from 1965 to 2020[3].

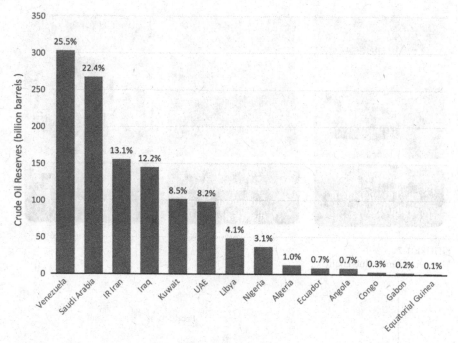

FIGURE 1.6
OPEC share of world crude oil reserves in 2018[4].

caverns along the Gulf Coast. The caverns have a storage capacity of 714 million barrels of oil.

1.5 Life Cycle of Oil Field

1.5.1 Petroleum Value Chain

The petroleum value chain is divided into three parts: upstream, midstream, and downstream. Upstream involves exploration and production of crude oil while downstream is refining of crude oil. Midstream involves transportation of crude from upstream to downstream, through pipelines, trucks, and by rail. The petroleum value chain is illustrated in Figure 1.7. Companies that have all three parts of the value chain are referred to as integrated oil companies.

1.5.2 Oil Field Life Cycle

The oil field life cycle for upstream has four main parts: exploration, development, production, and abandonment. The different phases of the life cycle and their associated cash flow characteristics are illustrated in Figure 1.8.

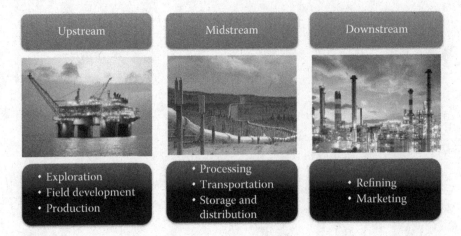

FIGURE 1.7
The petroleum value chain.

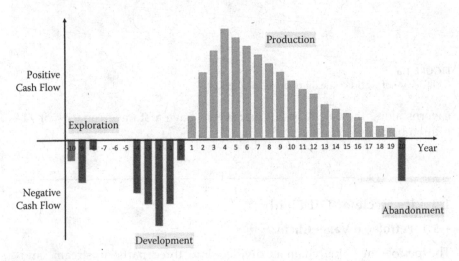

FIGURE 1.8
Oil field life cycle.

During exploration and development, capital costs are high leading to negative cash flow. Positive cash flows are realized during the production phase. Abandonment also has a negative cash flow.

1.5.2.1 Exploration

The life cycle of hydrocarbons begins with intelligent guesses about where they may be found. These guesses may be based on similarity of new sites with previous reservoirs, among other things. Seismic techniques are then used to map out the reservoir underground. This involves using sound waves and their reflections as they pass through different types of rock underground at different speeds. This is often done by setting off dynamite charges in specific patterns at different points above a suspected reservoir. The data collected are used to build a model of the reservoir to gain a better understanding of what the reservoir looks like underground. Simulations can be run to determine possible outcomes of locations of wells.

If the seismic results are favorable and show that there are hydrocarbons present, the next step is to drill an exploration well. Usually, the exploration well is drilled in the middle of the reservoir, followed by two or three appraisal wells at different points near the suspected edge of the reservoir. If no hydrocarbons are recovered from the exploration well, then no appraisal wells are drilled. Samples of rock, water, and oil and/or gas are collected from the wells. Hydrocarbon production rates are measured, as are the temperature and pressure in the reservoir. Logging tools are used to further verify the thickness of the reservoir and the rock and fluid types.

All of this information is fed back into the reservoir models and used to determine the type of hydrocarbons present, the quantity of hydrocarbons present, and whether the methods needed to recover them would be economic. If the project is determined to be economic, then the Development phase can be initiated. Exploration can last for many years, even decades.

Geoscientists and reservoir engineers (REs) are the primary practitioners involved in this phase, with input from drilling engineers and completions engineers.

1.5.2.2 Development

In the development phase, a plan is developed to maximize hydrocarbon production from the reservoir while minimizing costs. Drilling plans are developed taking into account the depth of the reservoir, the rock type, and the fluid type. Completions plans are also developed and these entail fracking in some reservoirs or managing sand from the reservoir in others. Processing facilities have to be designed to separate oil and/or gas from impurities such as water and carbon dioxide before the hydrocarbons can be sent on to the customer.

Different economic scenarios are run to ensure that the company's minimum financial metrics can be met over the life of the well at the company's discount rate, e.g., rate of return, net present value, and break-even point.

Geoscientists, REs, drilling engineers, completions engineers, and facilities engineers are the main practitioners involved in Development.

1.5.2.3 Production

Production may be considered the operations and maintenance phase. During Production, most of the reservoir has been drilled and the wells are consistently producing hydrocarbons. The main goal at this stage is to ensure the wells keep on producing so that they keep making money. Production is generally the only stage where a profit is made. If a well goes down, every effort is made to determine the cause and ensure that the cost to repair the well is not more than the expected profit to be made on the hydrocarbons remaining in the reservoir that the well can produce. After the hydrocarbons are produced, they must be processed to get to the customer's specifications and remove impurities such as carbon dioxide, hydrogen sulfide, and water. The hydrocarbons can then be sent to a refinery or processing plant. The Production phase can last from 6 to 60 years, depending on the reservoir type.

Production engineers are the main practitioners involved in this phase but they are greatly assisted by REs and geoscientists. Facilities engineers are responsible for processing of the recovered hydrocarbons.

1.5.2.4 Abandonment

Once a well has produced all that it can, it is plugged and abandoned. This process entails pumping cement into the well to plug it and prevent any fluids from coming up from the reservoir to the surface. In some cases, such as with offshore wells, decommissioning is carried out which includes removal of structures associated with the well such as platforms and pipelines. Abandonment can also involve returning the area to its original state.

The cost to decommission a well can be significant, especially if a lot of wells in a field reach the end of their lives at the same time. Sometimes companies sell off older reservoirs to pass on the cost of decommissioning wells to the buyer.

Production engineers are primarily responsible for abandonment with input from REs.

The various engineering disciplines, and geology, and the phases they are involved in are illustrated in Figure 1.9. The figure makes it apparent that

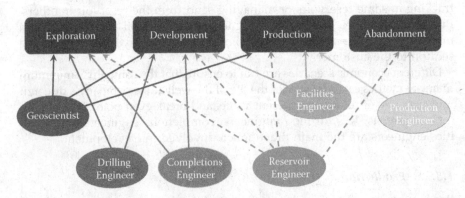

FIGURE 1.9
Engineering disciplines and their related phases in the hydrocarbon production life cycle.

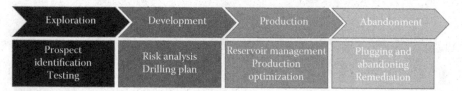

FIGURE 1.10
Phases of oil life cycle and associated activities.

REs are involved in every phase of the life cycle. REs are usually the project managers for the process, ensuring that information is translated and shared between disciplines as needed (Figure 1.10).

1.6 Engineering Disciplines in Oil and Gas Industry

The oil field life cycle is complex and requires the coordination of multiple disciplines. While this textbook focuses on engineering disciplines, it should be noted that the process also involves ensuring regulations are adhered to, legal agreements are signed, landowners are compensated, roads are built into new areas, economic analyses are run, and so forth.

Petroleum engineers are specifically trained to work in the oil and gas industry. Other engineering professionals which tend to be hired in the oil and gas industry include mechanical engineers, chemical engineers, civil engineers, and electrical engineers.

Notes

1 *What Drives Crude Oil Prices?* US Energy Information Administration, https://www.eia.gov/finance/markets/crudeoil/demand-nonoecd.php
2 *What Countries Are the Top Producers and Consumers of Oil?* US Energy Information Administration, https://www.eia.gov/tools/faqs/faq.php?id=709&t=6
3 *Statistical Review of World Energy,* BP, https://www.bp.com/en/global/corporate/energy-economics/statistical-review-of-world-energy.html
4 *OPEC Share of World Crude Oil Reserves,* OPEC, https://www.opec.org/opec_web/en/data_graphs/330.htm

2

Exploration and Geology

2.1 Exploration and Geology

The main task of geologists is to estimate the amount of crude oil or natural gas in a reservoir, whether the reservoir is unexplored or has producing wells.

As explained in Chapter 1, the oil field life cycle begins with the exploration phase when an oil or gas field has not yet had any production. The key questions to be answered during exploration are whether there are economic quantities of hydrocarbons present, where the hydrocarbons are located, and the type and quality of hydrocarbons.

As we cannot physically see underground, we have to make intelligent estimations of what lies beneath our feet. This is done by mapping the layers of rock primarily using seismic surveying. Seismic surveying uses the different speeds at which sound waves travel through different rock and fluid types to create maps of the subsurface and identify the potential location of hydrocarbons. Geologists are trained to read seismic maps like the one in Figure 2.1. The red and blue bands show different rock types layered upon each other. The y-axis represents the depth into the ground with the zero point being the surface. The x-axis is the areal extent of the surveyed area. Fault lines like the ones shown in yellow interrupt the flow of the layers and must be taken into account when drilling wells, or when planning locations for new wells. They may create pockets that trap hydrocarbons. Alternatively, they may create sinks that steal drilling fluids and compromise the process of drilling a well.

Seismic air guns, exploding dynamite, or vibrating trucks are used to generate sound. Reflected and refracted sound waves are detected by a receiver as shown in Figure 2.2.

Seismic surveying may be two dimensional (2D), three dimensional (3D), or four dimensional (4D). In 2D seismic surveying, both the sound source and the receivers are moved along a straight line. The result is a 2D image of the subsurface along the line. In 3D seismic surveying, the receivers are spread out in a dense array to pick up the reflected sound. The source is moved to a different location, with the receivers picking up the reflected

DOI: 10.1201/9781003100461-2

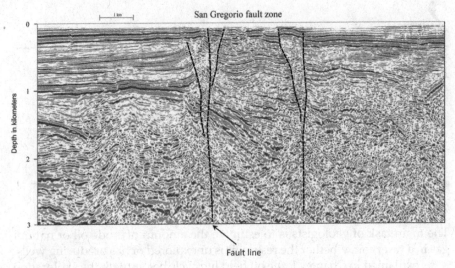

FIGURE 2.1
Two-dimensional seismic cross-section of the San Gregorio fault zone from the surface down to a depth of about 3 km[1].

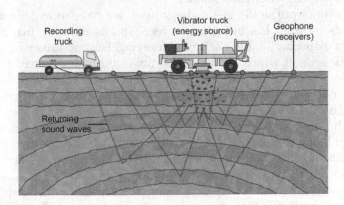

FIGURE 2.2
Seismic surveying.

and refracted sound from the new location. The resulting image is a 3D image of the subsurface.

A 4D seismic survey is carried out over the same area as a 3D seismic survey but at different times. The surveys may be carried out several years apart. This is done to track the changes in reservoir pressure, as well as changes in fluids and fluid saturations in the reservoir with hydrocarbon production. These observations may be used to optimize the placement of wells to maximize production, to identify drained and undrained areas of the reservoir, to monitor sequestration of CO_2, and to differentiate between fluid production and pressure changes in the reservoir.

2D seismic is cheapest and generally has less impact on the environment as fewer lines are needed. It is most often used for unexplored areas. Because 3D seismic provides greater resolution, it is used when hydrocarbons are known to be present. It is generally more expensive as it requires more hardware on the ground, as well as more computing power (Figure 2.3).

Petroleum is formed in source rock at high temperatures and pressures from the remains of prehistoric plants and animals. Higher temperatures resulted in lighter oils or natural gas forming. Petroleum migrates from the source rock into pores in reservoir rock until it is trapped and prevented from further migration by cap rock which is impermeable. Traps may be made of impermeable rock or salt. In fact, salt domes are characteristic signs in seismic that hydrocarbons may be present.

To be a reservoir, a rock must be porous, and the pores must be connected, i.e., the rock must be permeable or have high permeability. Hydrocarbons are generated in source rocks and then they migrate to reservoir rocks. Reservoir rocks store and transmit hydrocarbons. Sealing rocks trap hydrocarbons and prevent them from migrating from the reservoir (Figure 2.4).

There are four main types of traps as illustrated in Figure 2.5. The anticline trap has a seal at the top preventing hydrocarbons from migrating further upwards. In a fault trap, the fault displaces reservoir rock and prevents hydrocarbons from migrating. In a stratigraphic pinch-out trap, hydrocarbons that have migrated to the tip of the rock cannot migrate further due to the reservoir tip being surrounded by seal rock. In the

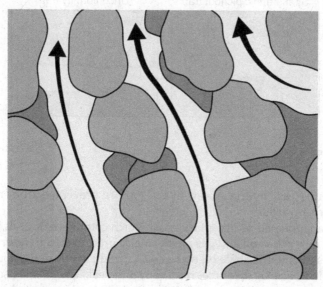

FIGURE 2.3
Permeability is connected pores, which provide pathways for fluid flow.

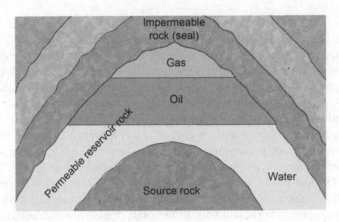

FIGURE 2.4
Schematic of source rock, reservoir rock, and trap.

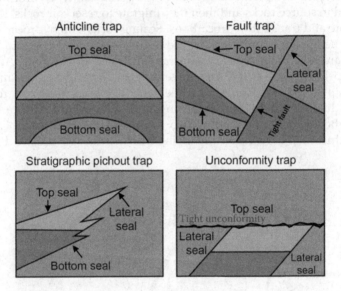

FIGURE 2.5
Different types of traps.

unconformity trap, hydrocarbons at the top are prevented from migrating laterally or upwards.

Shale is a laminated sedimentary rock, which comprises fine silt, clay, and other minerals. While oil and gas are normally produced from reservoir rock, in shale reservoirs, the shale oil and gas are produced from the source rock. The shale rock acts as a source, reservoir, and seal. This creates some challenges since source rock has very low permeability and is generally very thin.

Historically, petroleum was accessed by drilling vertical wells into the reservoir. This approach works well in conventional reservoirs but not so well in shale source rock which has only a thin section of rock containing petroleum. To maximize the amount of petroleum produced from shale, horizontal or directional drilling is used and combined with fracking.

2.2 Logging Techniques

Logging tools are used to measure resistivity (induction log, laterolog), radioactivity (neutron, gamma-ray, density), and acoustic (sonic). Logging is done during or after drilling unlike seismic. Logging results can be overlaid on seismic to give a more complete picture of the reservoir (Figure 2.6).

Geoscientists get information about the reservoir and surrounding rock layers from a process called well logging. A well log is a detailed record of changing properties with depth in a wellbore. First, a well is drilled then measuring tools are run down the well to measure various properties of the rock and fluids. Sometimes, the measuring is done concurrently with drilling. Various types of well logs and the properties they measure are described below.

Gamma-ray log – Measures the variation naturally occurring gamma radiation with depth to determine different rock types and rock thicknesses.

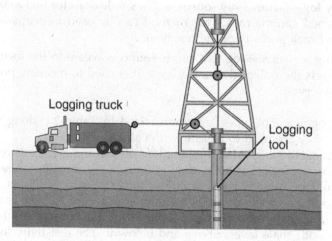

FIGURE 2.6
Schematic of the well-logging process.

Spontaneous potential (SP) log – Measures depth variation of electric potential difference between an electrode in the wellbore and an electrode at the surface. SP logs give information about rock type and thickness.

Core analysis – Analysis of a sample of rock taken from the reservoir during or after drilling. Core analysis can be used to valuate rock types, rock properties, and fluid properties. It can also be used to calibrate seismic and well-log measurements.

Mud log – A well log created by examining the drill cuttings circulated by drilling mud to the surface during drilling of a well. The cuttings provide information on rock types, fluid types, and fluid saturations.

Wireline formation tester – Measure pressures in the wellbore by pushing a probe into the formation and allowing production of fluids into a closed chamber. This method can provide information on fluid types, depths where different fluids contact each other, and permeability.

Electric log or resistivity log – Characterizes rock and fluid types by measuring the electrical resistivity of rocks and fluids. Hydrocarbons have a high electrical resistivity whereas formation water has a low electrical resistivity, i.e., a high electrical conductivity.

Neutron log – Uses a radioactive source to send high-energy neutrons into the formation and detects the energy of the neutrons in the formation. These measurements can be used to evaluate the porosity of the formation.

Density log – Gamma-ray source with a single detector that measures unabsorbed gamma rays. This method can be used to evaluate rock type and rock porosity of the reservoir.

Sonic log – Acoustic log that sends sound waves into the formation and detects the reflected sound waves. It is used to measure porosity of the reservoir.

Different logging tool types are summarized in Table 2.1 along with the properties they are used to measure or evaluate.

A triple combo log is a combination of gamma-ray, resistivity, and density-neutron logs. Together, they give basic information about the reservoir: the type of rock, types of fluids, and depths of the fluids.

An example of a triple combo log is shown in Figure 2.7. The gamma-ray log shows that there is a sand layer about 290-ft thick from a depth of 7,060 to 7,250 ft with shale layers above and below it. The resistivity log shows that hydrocarbons are present from 7,060 to 7,180 ft, which coincides with the top portion of the sand layer. At the bottom of the sand layer, below the hydrocarbons, is a brine aquifer from 7,190 to 7,290 ft. The density-neutron

TABLE 2.1

Logging techniques and the reservoir properties they measure

Reservoir property	Tool
Gamma-ray log	Rock type, rock thickness
Spontaneous potential (SP) log	Rock type, rock thickness
Core analysis	Rock type, rock properties, fluid properties
Mud log	Rock type, fluid type, fluid saturation
Wireline formation tester	Fluid types, fluid depths, rock permeability
Resistivity log	Rock type, fluid type
Neutron log	Rock porosity
Density log	Rock type, rock porosity
Sonic log	Rock porosity

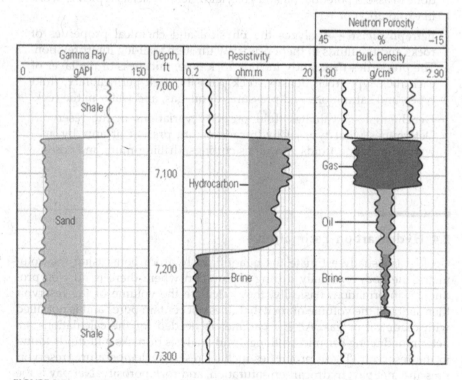

FIGURE 2.7
Triple combo log showing reservoir rock type, fluid types, and fluid depths[2].

log confirms the depths of the brine aquifer. It also differentiates between gas and oil in the hydrocarbon layer. It shows that gas can be found from 7,060 to 7,110 ft, and oil can be found from 7,110 to 7,190 ft.

2.3 Geoscience Disciplines

Geoscience is the study of the earth. In the oil and gas industry, geoscientists analyze geophysical, geological, and geochemical data to develop models of the subsurface and identify areas where oil and gas may be found. Geoscientists are able to use multiple techniques to develop an understanding of hydrogen generation, migration, and accumulation in a particular reservoir. This knowledge is used to reduce risk and make better exploration and development decisions.

There are many types of geoscientists with distinct specialties. They include the following:

Geophysicist – Determines trap size through remote sensing methods including seismic, gravity, magnetic, and electrical methods. This is done to assess potential oil and gas yield. Geophysicists typically work in exploration or research.

Petrophysicist – Analyzes the physical and chemical properties of rocks and the fluids in them primarily through well-log interpretation. The properties they study include rock type, thickness of the layer of each rock type, rock density, rock porosity, rock permeability, fluid types, and saturations and pressures of oil, gas, and water in the rock.

Geochemist – Evaluates fluid property variations in the reservoir. Determines what type of hydrocarbons are present, if any, by analyzing formation fluids as well as cuttings, drilling mud, and cores.

2.4 Hydrocarbon Estimation

Geologists usually estimate the quantity of hydrocarbons using the volumetric method, especially during exploration where there is no well production information. Basically, they calculate the volume of the reservoir then subtract the volume of rock that does not contain pores and is not filled with crude oil or gas. All the parameters needed for the calculations are determined using seismic, well logs, and analysis of rocks and fluids found in the reservoir. These properties include reservoir temperature, reservoir pressure, net pay, hydrocarbon saturation, and rock porosity. Net pay is the portion of the reservoir that contains hydrocarbons. Hydrocarbon saturation is the percentage of fluid in the reservoir that is hydrocarbons.

Figure 2.8 shows the workflow used by geologists to determine the quantity of hydrocarbons initially in the reservoir before production begins, i.e., the original hydrocarbons in place (OHIP). This could refer to original oil in place

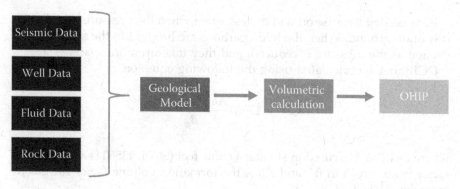

FIGURE 2.8
Workflow for estimation of original hydrocarbons in place by geologists.

(OOIP) or original gas in place (OGIP). All data from seismic, fluid and rock samples, and well logs are combined to create a geological model of the reservoir. This is an approximate representation of the reservoir built on a computer program designed for this purpose. The geological model is then used to calculate the quantity of hydrocarbons in the reservoir.

Volumetric estimation of OHIP requires knowledge of the following:

- Volume of the reservoir or subsurface rock that contains hydrocarbons. This includes the thickness of the reservoir as well as its areal extent.
- Average porosity of the reservoir.
- Hydrocarbon saturation, or the percentage of fluids in the reservoir that are hydrocarbons.

OOIP may be calculated using the following equation.

$$N = 7758Ah\Phi(1 - S_w)/B_{oi} \tag{2.1}$$

where

$$S_w = 1 - (S_o + S_g) \tag{2.2}$$

and N is OOIP in stock tank barrels (STB), 7758 is a conversion factor from acre-ft to barrels, A is the area of the reservoir (acres), h is the thickness of the reservoir which contains hydrocarbons (ft), Φ is the porosity of the reservoir rock (decimal), S_w is the water saturation (decimal), B_{oi} is the formation volume factor which gives the ratio of reservoir barrels to stock tank barrels or barrels at the surface (Reservoir BBl/STB), S_o is oil saturation in the reservoir (decimal), and S_g is the saturation of gas in the reservoir (decimal). If there is no gas present, then the value of S_g will be zero.

B_{oi} is needed because oil will fill less space when the pressure is higher as it is underground. When the hydrocarbons are brought to the surface, they expand as the pressure is reduced, and they take up more space.

OGIP may be calculated using the following equation.

$$OGIP = 43560\frac{Ah\phi(1 - S_w)}{B_{gi}} \tag{2.3}$$

where OGIP is measured in standard cubic feet (SCF), 43560 is a conversion factor from acre-ft to ft^3, and B_{gi} is the formation volume factor for gas in reservoir ft^3/SCF.

The volumetric estimate of hydrocarbons in place is an estimate at a moment in time. The hydrocarbons in place will change once the production of oil and/or gas begins. There are uncertainties in the value of OHIP due to the uncertainties in the factors that go into the calculation. For instance, the actual porosity of the reservoir varies at different parts of the reservoir, so using one average value could make the estimate more or less accurate.

The reservoir engineer takes the OHIP values and other relevant information such as the depth of the reservoir and shares them with the different engineering disciplines to move the project along, as described in the next chapter.

Notes

1 *Seismic Reflection Profile of San Gregorio Fault Zone*, USGS National Archive of Marine Seismic Surveys, https://www.usgs.gov/media/images/seismic-reflection-profile-0

2 Mark A. Andersen, Discovering the secrets of the earth, *Oilfield Review*, Spring 2011; 23(1): 60.

3

Reservoir Engineering

As explained in earlier chapters, exploration is the first phase in the oil and gas life cycle. A lot of the data collected in this phase is analyzed by geoscientists but some of it is analyzed by reservoir engineers as well. Reservoir engineers are involved in the entire oil and gas life cycle and tend to act as the project managers, especially during development. They are also the technical translators who ensure that the analyses carried out by the geoscientists can be reframed in engineering terms and understood by the other engineering disciplines involved in the life cycle. Finally, they are typically the asset managers and are tasked with ensuring that the oil and gas asset is optimized to maximize production and minimize costs.

3.1 How Oil and Gas Are Produced

Once the location and depth of oil and/or gas are established, wells can be drilled into the reservoir to produce the hydrocarbons. Because the pressure in the reservoir is high, when the hydrocarbons are brought to the surface, they expand. Figure 3.1 illustrates the possible outcomes of drilling a well: gas is produced from the gas cap, oil is produced, or no hydrocarbons are produced. If a well does not produce any hydrocarbons, it is referred to as a dry hole.

Before we dive into the roles and responsibilities of a reservoir engineer, let's take a look at the different types of reservoirs they have to work with. There are two main types of reservoirs: conventional and unconventional.

In normal, or conventional, reservoirs, hydrocarbons are produced from reservoir rock, and the hydrocarbons flow up the wellbore to the surface. The wells are usually drilled vertically into the reservoir. Unconventional reservoirs are further described in the following section.

3.2 Unconventional Reservoirs

Unconventional reservoirs are reservoirs that require special recovery techniques outside conventional operating practices. They may be characterized

DOI: 10.1201/9781003100461-3

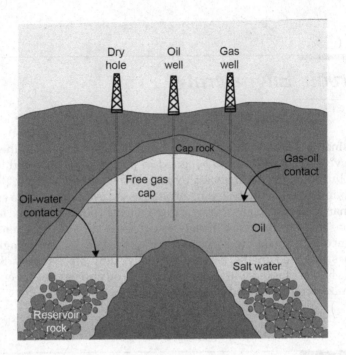

FIGURE 3.1
Possible outcomes of drilling a well.

by high viscosity crude oil, low permeability reservoirs, or production from source rock rather than reservoir rock. They require advanced technology to produce which makes them typically more expensive to produce than conventional reservoirs. The different types of unconventional reservoirs are described next (Figure 3.2).

3.2.1 Tight Gas

Low permeability reservoirs are referred to as tight. Tight gas is a hydrocarbon found in reservoirs with very low permeability. In order to produce tight gas at economic rates, artificial permeability must be created through fracking, i.e., fracturing the rock.

3.2.2 Shale Oil and Gas

The major shale plays or reservoirs in the United States are Permian, Bakken, Eagle Ford, Niobrara, Anadarko, Appalachia, and Haynesville. Appalachia includes the Utica and Marcellus shale plays. The locations of shale plays in the United States are shown in Figure 3.3.

Shale oil and gas are similar to tight gas except that shale oil and gas are found in source rock rather than reservoir rock. The source rock has very

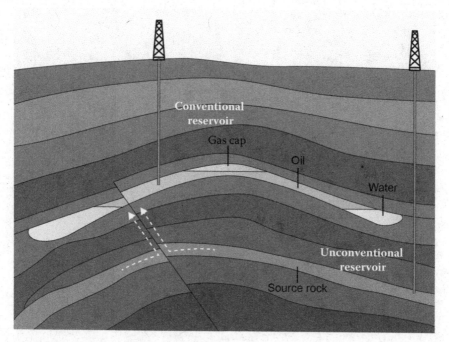

FIGURE 3.2
Conventional reservoir vs. shale unconventional reservoir.

low permeability, even lower than tight gas, so fracking is necessary. Additionally, because the shale rock is thin, horizontal wells are used to access more hydrocarbons than would be produced by a vertical well as is shown in Figure 3.4. The well is drilled vertically until just above the targeted rock. The well is then continuously drilled at an angle until it hits the targeted rock, then it is drilled horizontally.

3.2.3 Heavy Oil

Heavy oil is so thick that it is solid at room temperature and at reservoir temperature and pressure. It does not flow unless it is heated or mixed with lighter hydrocarbons. Steam is pumped into the reservoir to heat heavy oil enough for it to flow. In some locations, the reservoirs are close to the surface, around 200 feet, so steam recovery is not possible as it could lead to the surface collapsing as pressure support from oil is lost due to production. The heavy oil is mined instead using traditional mining methods whereby earth is excavated, mixed with water and sent to facilities where bitumen is separated.

One method of using steam to recover heavy oil is called Steam-Assisted Gravity Drainage (SAGD). This process is illustrated in Figure 3.5. Two

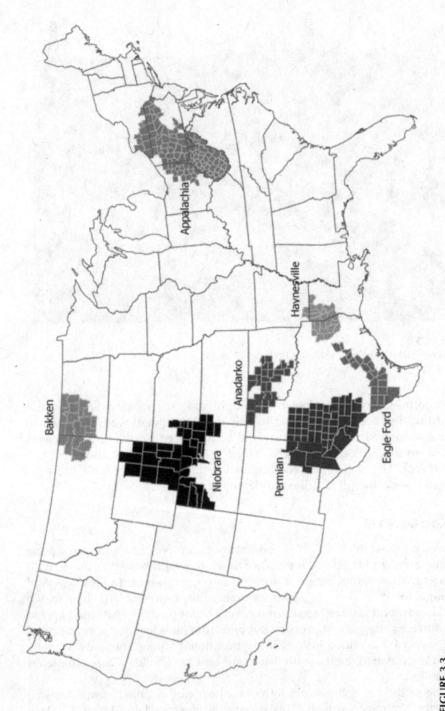

FIGURE 3.3
Map of shale reservoirs in the United States[1].

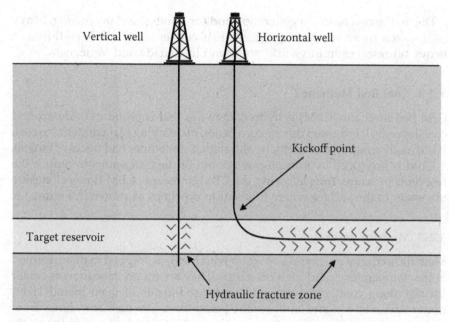

FIGURE 3.4
Area accessed by a fracked vertical well vs. a fracked horizontal well.

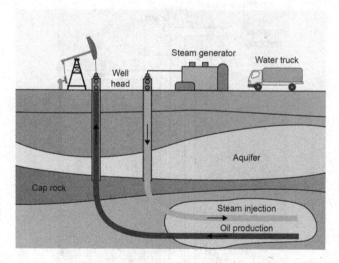

FIGURE 3.5
Recovery of heavy oil using steam-assisted gravity drainage.

horizontal wells are drilled into the reservoir. Steam is pumped into the reservoir through one well, and it returns to the surface through a second well. Eventually, the viscosity of the heavy oil reduces enough for it to flow and it is brought to the surface through the producer well.

Due to the cost of steam generation and/or mining, and processing, heavy oil is often more costly to produce than conventional oil. Two-thirds of heavy oil reserves in the world are found in Canada and Venezuela.

3.2.4 Coal Bed Methane

Coal Bed Methane (CBM) is hydrocarbon gas that is found in underground coal deposits. It is formed during conversion of decaying plant material to coal. It is usually composed of methane although it sometimes had traces of ethane.

CBM is produced by pumping water out of the coal seams to reduce the reservoir pressure, thus allowing the CBM to escape. CBM flows alongside the water to the surface where it is sent to pipelines as shown in Figure 3.6.

3.2.5 Methane Hydrates

Methane hydrates are crystals of water with methane trapped in them, formed at low temperatures and high pressures. They are found in sediments under the sea along continental margins, under sediments of deep inland lakes, under Antarctic ice, and in the arctic permafrost. The methane comes from biological activity in sediments, or geothermal activity deep within the earth.

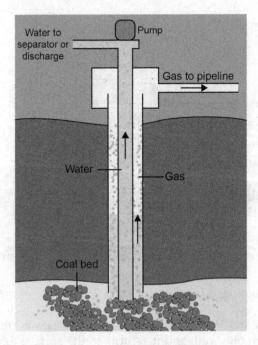

FIGURE 3.6
How coal bed methane is produced.

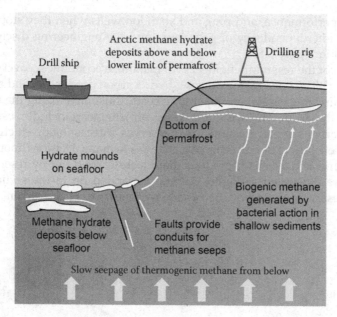

FIGURE 3.7
Types of methane hydrate deposits.

Methane hydrate comprises a methane molecule trapped in a solid lattice of water molecules. One cubic meter of gas hydrate (35 cubic feet) releases 164 cubic meters (5,791 cubic feet) of methane at surface conditions.

The US Geologic Society estimated in 2008 that the North Slope of Alaska has between 25.2 and 157.8 trillion cubic feet of methane in methane hydrates. Methane hydrates cannot currently be recovered in economic quantities. Research continues into commercially viable methods to produce methane hydrates. However, methane hydrates have proved difficult to study because they melt when they are brought to the surface (Figure 3.7).

3.3 What is a Reservoir Engineer?

Reservoir engineers are the project managers of the extraction of oil and gas from the ground. They are responsible for taking geologic information from the geoscientists, converting it to engineering terms, picking locations for new wells, getting costs from drilling and completions engineers, running analyses to determine if the new well would be economic. They also share information on hydrocarbons and other fluids in the reservoir with facilities engineers who design separations facilities. Finally, they work with production engineers to restore production when wells go down, monitor reservoir performance, and plug and abandon wells when they stop producing. The interdependency of geoscience and the engineering disciplines is illustrated in Figure 3.8.

The role of the reservoir engineer is to maximize economic recovery from a reservoir. This is done by maximizing hydrocarbon production and recovery while minimizing capital and operating costs, thus maximizing the ultimate value of the field. The role of a reservoir engineer includes estimation of hydrocarbons in place, estimation of reserves, forecasting production of oil and gas wells, and running economic analyses to make decisions about which wells to drill and when. These four main tasks carried out by reservoir engineers are illustrated in Figure 3.9 and expanded on in the sections that follow. Reservoir engineers often create workflows to reduce errors.

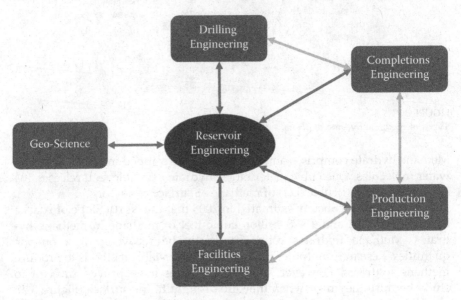

FIGURE 3.8
Interdependency of reservoir engineering and geoscience and other engineering disciplines.

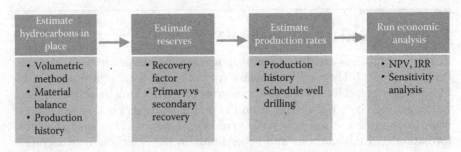

FIGURE 3.9
Steps in the reservoir engineering process.

3.4 Estimation of Hydrocarbons in Place

Reservoir engineers work very closely with geoscientists to estimate the original quantity of hydrocarbons present in a reservoir before any hydrocarbons are extracted. This value is called Original Hydrocarbon In Place (OHIP). It is important to know the OHIP so that plans can be made for how to extract the reservoir, and a determination can be made of whether it is economical to extract from the reservoir. Note that hydrocarbons in place refers to whatever hydrocarbons are in the ground at the time of estimation. This value can be determined even after production has begun.

Estimation of hydrocarbons in place can be done through various methods as described below. Pressure in the reservoir is different from pressure at the surface. Therefore, hydrocarbon volumes must be converted to and expressed in surface conditions.

- **Analogy** – This method uses performance characteristics of reservoirs believed to be similar to the reservoir under study. This method is mostly used for a newly discovered or poorly defined reservoir when no factual data on fluid or reservoir properties is available for the reservoir. The already producing reservoir should be close to abandonment so that its resources and recovery factors are known. Analogy can be used to determine the order of magnitude of the hydrocarbon resources, as well as how much of the hydrocarbons can be produced, i.e., the reserves.

- **Volumetric analysis** – This is the simplest method to determine hydrocarbons in place. It is a static measure of hydrocarbons in place, i.e., it has no time dependency. It is generally used in the early stages of reservoir development as it requires no production data. First, the overall volume of the reservoir is determined using the reservoir boundaries and thickness. The porosity of the rock and the saturation of hydrocarbons is then used to calculate the volume of hydrocarbons in the reservoir. Reservoir volumes are converted to surface volumes using a conversion factor called the formation volume factor.

- **Material balance** – This method is based on the law of conservation of mass. It uses analysis of pressure behavior as fluids are withdrawn from the reservoir. It uses fluid production data as well as rock and fluid properties. This method assumes that the reservoir is homogenous, fluid production occurs at a single point, fluid injection occurs at a single point, and the direction of flow in the reservoir is unimportant. Therefore, reservoir pressure declines proportionately to gas production.

3.5 Estimation of Reserves

Resources are the total amount of hydrocarbons estimated to be present in a reservoir, while reserves refer to the amount of oil or natural gas that can be economically produced at current prices. It depends on recovery efficiency as well as economics of operation. Therefore, reserves can increase or decrease as operating expenditures, capital expenditures, and prices of oil and gas change.

In conventional oil reservoirs, on average, only about 20%–30% of oil can be recovered due to various limitations. In conventional gas reservoirs, about 80%–85% of the gas in the reservoir can be produced. These percentages are referred to as the recovery factor (RF). In order to calculate the recoverable reserves, the RF must be known.

3.5.1 Recovery Factor

Recovery Factor (RF) is the percentage of original hydrocarbons in place that can be economically produced. RF depends on the reservoir rock type, the hydrocarbon type, the displacement fluid, and the shape and extent of the reservoir. RF can be calculated using Eq. 3.1.

$$\text{Recovery Factor} = \frac{\text{Estimate of recoverable hydrocarbons}}{\text{Estimate of hydrocarbons in place}} \quad (3.1)$$

For oil reservoirs, RF is usually low, around 20%. To recover more from the reservoir, secondary and tertiary recovery methods are employed (Figure 3.10).

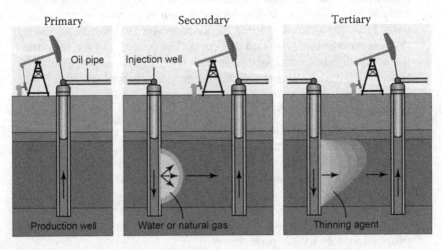

FIGURE 3.10
Driving force for oil recovery in primary, secondary, and tertiary recovery.

3.5.1.1 Primary recovery

Primary recovery refers to the natural production from the reservoir whereby oil flows into the well under its own natural pressure. It is also called pressure depletion drive because it relies on the pressure in the reservoir declining. The use of pump jacks and other artificial lift methods also falls under primary recovery. Drive mechanisms are the natural energy of the reservoir that can be used to drive hydrocarbons into the wellbore. By characterizing the drive mechanism of a specific reservoir early on, reservoir engineers can maximize the recovery of hydrocarbons. Fluid production ratios, i.e., the relative amounts of water and oil or gas produced, and reservoir pressures can be used to determine drive mechanisms (Table 3.1).

- **Solution gas drive:** As oil is removed from the reservoir, the pressure in the reservoir drops, causing the remaining oil and the gas dissolved in it to expand. This expansion provides most of the drive energy for the reservoir.

- **Expansion drive:** Expansion of rock and water as pressure falls in the reservoir provides additional drive energy.

- **Gas cap drive:** A gas cap is initially present in the reservoir. As oil is produced from the reservoir, the pressure drops, leading to expansion of the gas cap which provides drive energy. The larger the gas cap, the slower the pressure decline. Expansion of the oil also provides drive energy hence oil viscosity has to be considered when determining the recovery factor.

- **Water drive:** There is an aquifer in contact with the oil in the reservoir. Expansion of the water in the aquifer provides energy that drives the production of oil. In a bottom water drive, the aquifer is located below the reservoir. In an edge drive, the aquifer is located on the edge of the reservoir. The two water drives are illustrated in Figure 3.11. The bigger the reservoir and the greater the permeability, the greater the water drive and the stronger the energy provided by the aquifer.

TABLE 3.1

Primary drives and their associated oil recovery factors

Primary drive mechanism	Recovery Factor (%)
Solution gas	5–30
Expansion	2–5
Gas cap drive	20–40
Water drive – Bottom	20–40
Water drive – Edge	35–60
Gravity drive	5–30 (incremental)

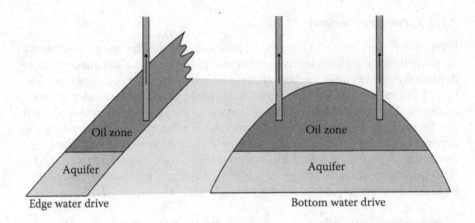

FIGURE 3.11
Location of the aquifer in edge water drive and bottom water drive.

- **Gravity drive:** The energy comes from difference in density of the fluids in the reservoir. Gravity drive is present in reservoirs with a gas cap or water drive. The energy comes from liberated gas moving upwards into the gas cap, or water moving downwards into the aquifer. The moving gas pushes oil toward the well.

- **Combination:** It is usual for a reservoir to be influenced by more than one type of drive. This is referred to as a combination drive. For instance, a reservoir may have both an aquifer and a gas cap. However, the dominant drive mechanism will determine production and recovery trends.

3.5.1.2 Secondary recovery

Secondary recovery refers to methods used to produce oil after natural recovery methods have been exhausted. It is primarily water flooding, and in rare cases, natural gas injection. These methods increase reservoir pressure so that more oil can be recovered. Secondary recovery can increase oil recovery to as much as 45%.

Water flooding is the most common secondary recovery method. It involves pumping water into injection wells to displace oil towards the production wells. Sometimes, water is replaced with other fluids such as carbon dioxide.

3.5.1.3 Tertiary recovery

Tertiary recovery refers to removal of residual oil that has not been produced through primary or secondary recovery methods. It is also called Enhanced Oil Recovery (EOR). EOR can increase oil recovery up to about

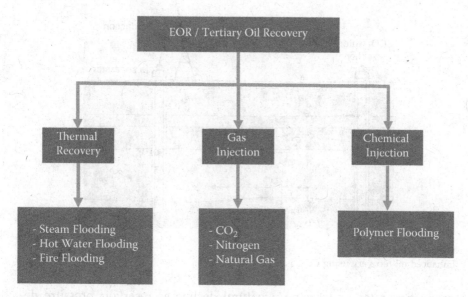

FIGURE 3.12
Tertiary oil recovery methods.

65%. Tertiary recovery restores formation pressure through the following methods (Figure 3.12).

Thermal Recovery: Heat is introduced into the reservoir via steam injection, hot water injection, or combustion, to reduce the viscosity of heavy oil so that it can be pumped out of the reservoir. One example of this method is called SAGD which was described in Section 3.2.3.

Gas Injection: Injecting gases such as carbon dioxide (CO_2). The gases diffuse into oil in the reservoir, reducing its viscosity and causing it to expand thus increasing its flow towards the producing wells. Carbon dioxide is removed from the produced oil and recycled back into the reservoir. Some CO_2 remains sequestered in the reservoir. Other gases such as nitrogen or natural gas may be used for EOR, though much less frequently than CO_2 (Figure 3.13).

Chemical Injection: Injection of polymers into the reservoir. These polymers are surfactants and work by reducing surface tension of oil, making it easier for water to sweep oil toward the producing wells. Polymers also increase the viscosity of water, increasing its efficiency in sweeping the remaining oil from the reservoir into the producing well in a process called polymer flooding.

3.5.2 Production History

Production history is an analysis of production behavior as fluids are withdrawn from the reservoir. It is used in later stages of development with

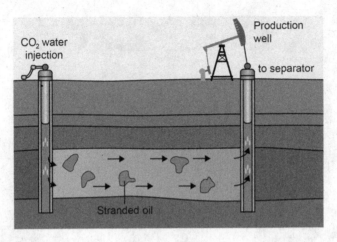

FIGURE 3.13
Enhanced oil recovery using CO_2.

production rates undergoing a natural decline as reservoir pressure de-creases. Several methods may be used to match equations to production history. These methods depend on reservoir characteristics and/or average production of wells in that part of the field or reservoir.

Once a good match is achieved between the equation and data, future production can be forecasted for the well. The Estimated Ultimate Recovery (EUR) of the well can then be calculated. EUR is the amount of hydrocarbon a well is expected to produce over its lifetime (Figure 3.14).

3.5.3 Reserves Classification

A company's hydrocarbon reserves provide an indication of the company's value. Reserve amounts are filed with the Securities and Exchange Commission (SEC). Estimated recoverable hydrocarbons may be classified as reserves if they are associated with a commercially viable project. Reserves can be categorized as either proved or unproved (Figure 3.15).

3.5.3.1 Proved Reserves

The term proved reserves, or proven reserves, refers to the quantity of hy-drocarbons that a company expects to be able to produce with 90% certainty from a given field from a technical and commercial standpoint (Figure 3.16). OPEC countries had 70.1% of the world's proved reserves at the end of 2019. Proved reserves can be further classified into Proved Undeveloped and Proved Developed. Proved Undeveloped reserves are those that are expected to be recovered from existing wells with major capital expenditure, or from new wells on undrilled acreage. Proved Developed reserves are expected to be recovered from existing wells with existing operating procedures.

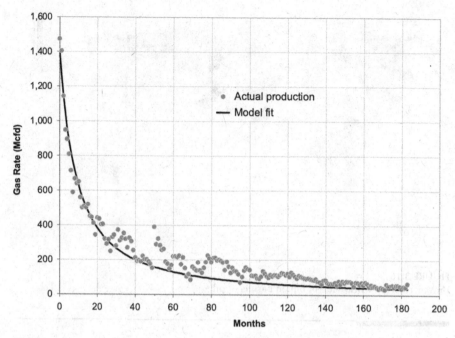

FIGURE 3.14
Gas production over time plot showing actual data and an equation fit to the data.

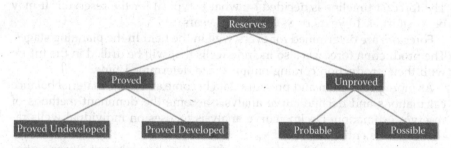

FIGURE 3.15
Reserves categorization.

3.5.3.2 Unproved Reserves

Unproved, or unproven, reserves are reserves that have a low certainty of being recovered. Unproved reserves may be further divided into Probable or Possible Reserves. Probable reserves have a 50% certainty of commercial extraction, and Possible reserves have a 10% certainty of commercial extraction. Anything below 10% is referred to as Contingent Resources.

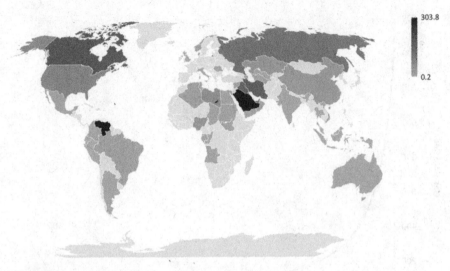

303.8

0.2

FIGURE 3.16
Proved reserves worldwide in billion barrels of oil.

3.6 Determination of Rates

Production forecasts for individual wells may be generated using the start rate for the well, and the typical decline rates for the reservoir under study. The forecast timeline is decided by what is typical for the reservoir. It may be as short as 10 years or as long as 30 years.

Forecasts are determined for every well in the field In the planning stages. The production forecasts also include wells that will be drilled in the future with their production coming online at the determined time.

As more production and pressure data become available, material balance calculations and decline curve analysis become the dominant methods of reserves estimation. Decline curve analysis focuses on individual wells rather than the entire reservoir.

Every parameter used in reserves estimation has inherent uncertainties which causes uncertainties in the final calculations (Figure 3.17). OHIP, EUR, reserves, and other factors may be calculated using the deterministic method or the probabilistic method. The deterministic method uses a single value of each parameter to arrive at a single answer, typically a mean or average value. Tornado charts may be used to compare the relative importance of the parameters. The probabilistic method is more rigorous and uses a distribution curve for each parameter, resulting in a range of values for the answer. Monte Carlo simulations can be used to generate the parameter distribution curves.

With the probabilistic method, only the end result is known while the exact parameter values are not. On the other hand, the deterministic method may ignore variability in the parameter data. Comparing calculations using

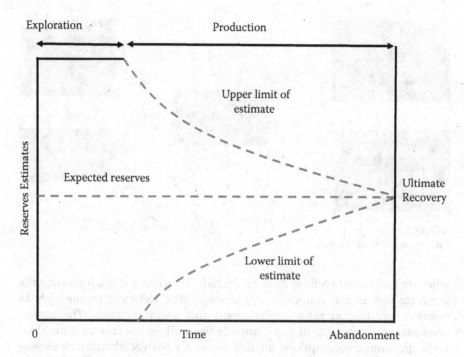

FIGURE 3.17
Uncertainty in reserves estimation.

both methods may provide confidence in the final results. If the answers are very different, the assumptions may need to be revisited.

In later stages, the required information includes locations of missed opportunities, whether the reservoir is continuous or compartmentalized, and whether production equipment is functioning properly. Reservoir engineers determine whether enhanced recovery techniques are working properly and if reservoirs are fully tapped.

Different scenarios need to be run for both the forecasting and the uncertainties and assumptions. Assumptions and sources of uncertainties include reservoir type, source of reservoir energy, geological, engineering, and geophysical data. The experience and knowledge of the evaluators also play a part. Two engineers could look at the same data and come to slightly or very different conclusions.

3.7 Economic Analysis

The production forecasts are used to run economics analyses to determine profitability of the field. Economic analyses are also run to determine

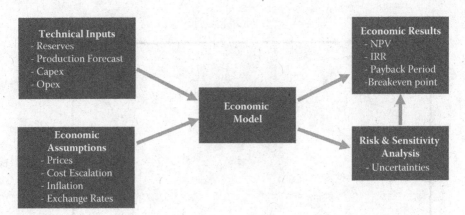

FIGURE 3.18
Parts of an economic analysis.

whether individual wells should be drilled. An economic analysis requires technical inputs and economic assumptions. The technical inputs include reserves, production rates, capital costs, and operating costs. The capital costs include costs to drill and complete the well, in the case of individual wells. Economic assumptions include prices for hydrocarbons, cost escalations due to inflation and other factors, and taxes. Risk and sensitivity analyses are run to evaluate the impact of the variation of the inputs and assumptions. The basic outputs of the economic analysis are the Net Present Value, the Internal Rate of Return, and the payback period (Figure 3.18).

Because of inflation, any money received in the future will be worth less than if it was received in the present. The Net Present Value (NPV) is the sum of all future cash flows, negative and positive, converted to their equivalent value in the present at a specific discount rate. The discount rate is decided by each company as a hurdle which must be overcome for a project to be deemed profitable. If the NPV of a project is positive, then the project is attractive (Figure 3.19).

The Internal Rate of Return (IRR) is the discount rate at which NPV equals zero. It is essentially the rate at which the project breaks even. The higher the IRR, the more desirable the project is for investment. If the IRR is higher than the company's hurdle rate, then it is deemed worthwhile to pursue.

The payback period is the amount of time it takes for a project to recover its investment costs (Figure 3.19). The shorter the payback period, the more attractive the project.

The level of uncertainty decreases as more data becomes available (Figure 3.20). Thus, uncertainty is highest during exploration, and lowest during the production phase. Types of uncertainty change with each phase. In the development phase, the main uncertainties are in the project execution, e.g., in the reservoir and fluid properties, facilities, schedule, and budget. During production, the main uncertainties are more technical such

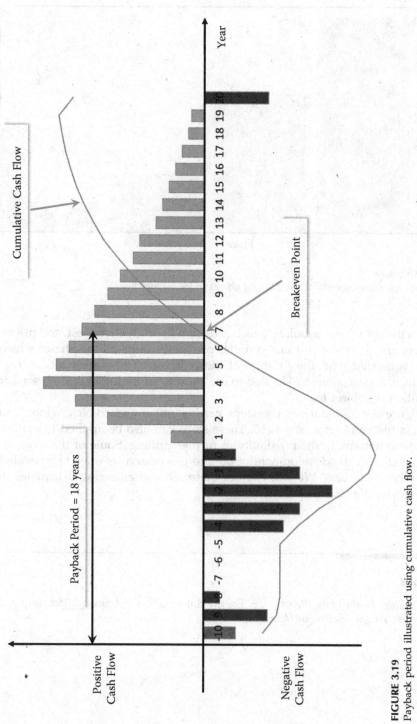

FIGURE 3.19
Payback period illustrated using cumulative cash flow.

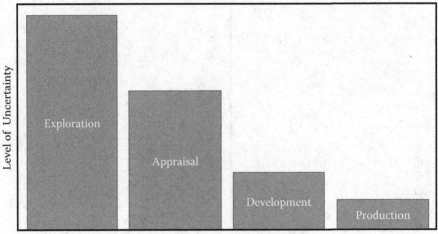

FIGURE 3.20
Relative levels of uncertainty at different phases of the oil life cycle.

as in the operating cash flow which is related to production, cost, and prices. There are also risks and uncertainties present through all the phases which are associated with the location of the reservoir. These are political, regulatory, and economic risks tied to the country or region that the reservoir location is subject to.

Figure 3.9 summarizes the steps carried out by reservoir engineers to assess the value of a new field. These steps can also be applied to mature fields to determine the possibility of further drilling. Some of the costs associated with the development of oil and gas reservoirs must be provided by other engineers. We will explore these other engineering disciplines in the upcoming chapters.

Note

1 *Drilling Productivity Report*, US Energy Information Administration, https://www.eia.gov/petroleum/drilling/

4

Drilling Engineering

The drilling engineering discipline is involved with drilling wells to access hydrocarbons that are underground. It involves the planning, designing, and costing of wells, as well as planning and implementation of well testing. Drilling engineers will design a drill job for a new well based on information provided to them by the reservoir engineer.

4.1 How a Well Is Drilled

Drilling is done to create a conduit into a hydrocarbon reservoir so that oil and/or gas may be sent to the surface. The complexity of the drill job varies depending on various factors including the type of reservoir, the depth of the reservoir, and the rock types that must be drilled through. Drilling in water, also called offshore drilling, is more complex because the depth and movement of water must be taken into consideration.

4.1.1 Onshore Drilling

A motorized drill bit is used to drill a hole in the ground 5–30 inches wide (Figure 4.1). The drill bit has teeth that can cut or crush rock. The drill is lubricated and cooled by a fluid called drilling mud. The drilling mud circulates into the hole and brings up bits of rock to the surface. These rock bits are called drill cuttings. Drilling mud also fills up the drilled hole, pressurizing it and preventing water and other fluids from flowing into the wellbore. Drilling mud may be oil-based or water-based. The drilling mud type is chosen to minimize damage to the formation being drilled. Oil-based muds generally use diesel as a base. However, oil-based muds may interfere with geochemical analysis of cuttings as the analysis cannot distinguish between diesel in the mud and crude oil from the formation.

When drilling mud comes back to the surface, it is sent to a reserve pit where the fine particles picked up during drilling are allowed to settle to the bottom. The mud is recirculated back to the well. Various chemicals are added to the mud to eliminate bacteria, reduce fluid loss, increase the viscosity of the mud, and enhance mixing and diffusion, among other functions.

The well is drilled to a depth just below where the reservoir is thought to be. A rotary rig can drill up to 1,000 feet per day. During the drilling

DOI: 10.1201/9781003100461-4

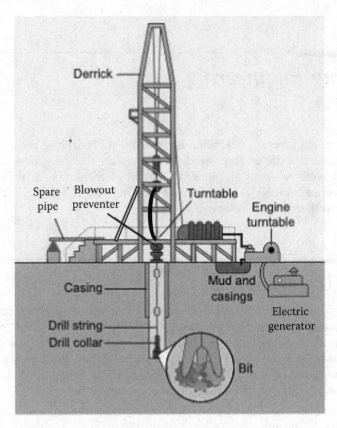

FIGURE 4.1
Components of an onshore drilling rig.

process, well logging is carried out to get information about the rocks and fluids in the reservoir and surrounding rock layers.

4.1.2 Offshore Drilling

To reach reservoirs below the sea, offshore rigs are employed. They tend to be floating rigs in water depths over 400 meters because at those depths, fixing a rig to the sea floor is very costly. Once drilling is complete, the floating drill platform moves on to the next drilling location. Sometimes, drilling ships are used instead.

Drillships are used in water depths from 2,000 to 12,000 feet. They contain drilling equipment and can move themselves from one drill site to another. In shallower waters, drillships are anchored to the sea floor while drilling. In deeper waters, however, drillships must rely on dynamic positioning systems to stay in place (Figure 4.2).

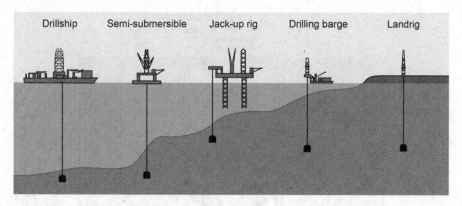

| Drillship | Semi-submersible | Jack-up rig | Drilling barge | Landrig |

FIGURE 4.2
Offshore and onshore drilling rigs.

Semi-submersible rigs are used at depths up to 10,000 feet. Unlike drill-ships, semi-submersibles must be towed from place to place by an outside transport vehicle. Jack-up rigs have self-elevating platforms with legs that can be lowered to the sea floor. The drilling equipment is jacked up on the platform above the water. The jack-up rig operates at depths up to 500 feet. It floats when the legs are not deployed. Some jack-up rigs can self-propel from site to site while others need to be towed by tug boats or submersible barges.

Drilling barges operate in shallow waters, around 500 feet, and calm conditions. They must be towed by tugboat to move from one location to another. Drilling equipment is placed on the deck and the barge is held in place by anchors during drilling.

4.2 Parts of a Well

The well is drilled to a pre-determined depth of several hundred or several thousand feet, and lined with a metal pipe called surface casing, to protect aquifers and act as support for production casing which protects the rest of the wellbore. Cement is pumped behind the surface casing to prevent fluids from moving into or out of the well.

A blowout preventer is attached to the surface casing. In the event that pressure in the wellbore becomes too high, the blowout preventer can be closed to prevent high-pressure gas or liquids from going out of the well. Production casing is perforated by completions engineers to allow fluids to flow into the wellbore. This will be explored further in the next chapter.

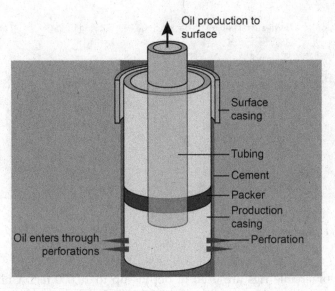

Oil production to
surface

Surface
casing

Tubing

Cement

Packer

Production
casing

Oil enters through
perforations

Perforation

FIGURE 4.3
Parts of a well.

A smaller diameter pipe, from 2.5 to 7 inches in diameter, called production tubing is run into the well inside the production casing. Hydrocarbons and other fluids are produced up the tubing to the wellhead (Figure 4.3).

4.3 Directional Drilling

Directional drilling is drilling wells at angles that deviate away from the vertical. It is done to access more oil and gas than could be reached using vertical wells. The shale revolution in the US was due to a combination of horizontal well drilling and fracking.

4.3.1 Horizontal Drilling

This type of drilling starts with a vertical well which turns horizontal as it nears the reservoir. The horizontal portion is often up to a mile long in order to reach as much of the reservoir as possible. The reservoir in these cases is often very thin but wide-spread. Horizontal wells allow oil to be produced in economic quantities even from thins reservoirs. They can produce up to 20 times as much oil and gas as vertical wells. They have the added benefit of requiring fewer wells which minimizes disturbances on the surface (Figure 4.4).

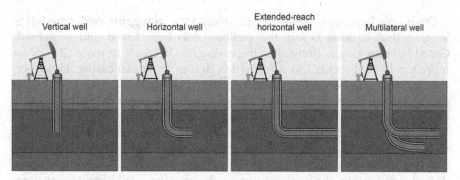

FIGURE 4.4
Directional drilling types and a vertical well for comparison.

4.3.2 Multilateral Drilling

Multilateral drilling involves drilling wells into reservoirs at different depths from the same main borehole. It allows accessing of different reservoirs from the same wellbore.

4.3.3 Extended Reach Drilling

This is drilling of wells that are relatively shallow vertically but extremely long horizontally. It is generally accepted that a well qualifies as extended reach if its horizontal length is more than twice its vertical depth. Extended reach wells are often used to reach distant reservoirs to reduce the impact of drilling directly over the reservoir. They are, however, challenging and expensive to drill.

4.3.4 Multiple-Well Pad Drilling

This is an application of directional drilling rather than a type of directional drilling. It involves drilling multiple wells in different directions from a single drill site. It saves money and allows oil to be produced more efficiently. It also minimizes the environmental and operational impact of drilling multiple wells on different sites.

4.4 Well Logging

Advances have been made in well logging such that logging and drilling can be carried out simultaneously. Historically, drilling had to be completed first before logging could be done.

Open hole logging is the most prominent type of logging performed. It is logging that is done before the wellbore has been cased and cemented. Cased hole logging is done after the casing has been set in place. It is done more rarely as it yields less information due to the metal casing. However, readings can still be useful and can provide information on the state of the casing, perforations, and cement, and potential blockages in the wellbore. It can also accommodate gamma-ray and neutron porosity logs.

Logging-While-Drilling (LWD) incorporates logging tools into the drill string so that logging occurs concurrently with drilling. These petrophysical data are sent to the surface in real time. Information can also be stored and downloaded when the tools return to the surface. LWD incorporates resistivity, mud logs, SP, induction, and gamma-ray tools. LWD allows drillers to make immediate changes to the drilling plan.

Measurement-While-Drilling (MWD) refers to information that is used to enhance drilling efficiency and wellbore geometry, such as direction and orientation. It is primarily used for directional drilling to ensure the well stays in the zone of interest. Information is usually relayed to the surface using pressure pulses in the mud in the form of positive, negative, or continuous sine waves.

4.5 Costs

There are several costs associated with drilling a well. Costs can vary depending on various factors including the depth of the reservoir, the rock types being drilled through, the fluid types, or the type of well that is being drilled, i.e., vertical, horizontal. The drilling engineer gets this information from the reservoir engineer and designs a drill job, then shares the cost with the reservoir engineer. The costs can include the daily cost of the drilling rig, materials including chemicals, drilling mud and water, casing, cement, fuel costs, and logging.

Offshore drilling costs more than onshore. Exploration wells are much more expensive than wells drilled during the development phase, as development wells can take advantage of economies of scale since many wells are being drilled at the same time. A typical horizontal well onshore can cost roughly $6 million, a vertical onshore well can cost $2 million, while an offshore well can cost $200 million.

After a well is drilled, it is handed over to the completions engineer to be completed. This is explained in the next chapter.

5

Completions Engineering

Completions is the process of preparing the well for production, and maximizing production of oil and gas from the surrounding reservoir rock.
Completions involves putting piping in the wellbore, cementing, perforating,
stimulating, installing sand control, and installing production equipment at
the wellhead.

Maximizing production is done by stimulating the well, and installing sand
controls to keep sand from the reservoir from clogging perforations in the
well. Completions are carried out after the well has been drilled and the well
has been cemented and casing put in place. They optimize the connection
with the reservoir and help determine the long-term health of the well.

A Christmas tree or production tree is a device placed at the surface at the
wellhead to regulate flow of fluids into pipelines that take hydrocarbons to
processing facilities. It can be used to shut down production from the well
for repairs. Some Christmas trees can be remotely monitored and operated.

5.1 Open Hole vs. Cased Hole

There are two types of completion: cased hole and open hole. If production
casing is not used, the well is said to be an open-hole completion. In an
open-hole completion, drilling is done using drilling mud that matches
conditions in the well in order to prevent the wellbore from collapsing in on
itself. After drilling is completed, production casing is inserted to maintain
the integrity of the well. However, the casing does not extend down to the
depth of the reservoir. No perforations are needed to allow fluid flow from
the reservoir into the wellbore.

Open-hole completions are cheaper than closed hole because in open hole,
no cementing or perforating operations are performed, both of which are
expensive especially in horizontal wells. However, it is difficult to control
production of sand and unwanted formation fluids from the reservoir since
there is no casing over the reservoir. It is difficult to carry out remedial work
on open-hole wells as the reservoir cannot be isolated.

In a cased-hole completion, surface casing and production casing are inserted into the wellbore and secured using cement to fill the annulus between the casing and the formation. The production casing covers the

DOI: 10.1201/9781003100461-5

reservoir and must be perforated to let reservoir fluids into the well. Perforations are carried out using perforation guns to set off controlled explosions which create holes through parts of production casing near the reservoir. Completions engineers help to select appropriate tubing size for desired flowrates. Fluids flow through the perforations into the production casing, up the tubing to the surface. Unlike casing, tubing is not cemented and can in some cases be pulled up for inspection or replacement.

Cased-hole completions allow multiple reservoirs or zones to be produced in a single well as the production casing helps to isolate the reservoirs. Remedial processes are also easier to perform. Casing enhances the integrity of the well and controls reservoir sand and fluid flow into the well. However, production casing could limit the volume of hydrocarbons produced from the well.

5.2 Stimulation Techniques

Selection of stimulation technology depends on well and reservoir characteristics. Some reservoirs have high permeability and do not require stimulation to get hydrocarbons to flow. Others are much less permeable and must be stimulated to get production of hydrocarbons.

Stimulation is used to induce or increase production in a well. In reservoirs with low permeability, stimulation initiates production from the reservoir. It may also be used to restore production to a well that has lost productivity. The two main types of stimulation are acidizing and fracking. Acidizing was used long before fracking was developed in the late 1940s.

5.2.1 Acidizing

Acidizing involves pumping acid into the well to improve the permeability of the reservoir. The acid dissolves limestone, calcite, and dolomite between the reservoir rocks. Acidizing requires the use of acid corrosion inhibitors to protect the steel casings and tubulars in the well. Design of an acidizing job depends on the type of formation rock, and the permeability of the formation. It is suitable for sandstone, shale, and carbonate formations. Dilute hydrochloric and hydrofluoric acids, between 1% and 30%, are the typical acids used for acidizing.

Matrix acidizing involves pumping relatively low-pressure acid into the well to dissolve the sediments that are inhibiting permeability, thus enlarging the natural pores and increasing the flow of hydrocarbons.

Fracture acidizing involves pumping high-pressure acid into the well. This creates physical fractures as well as dissolves sediments, forming channels that allow hydrocarbons to flow.

In acid washing, acid is used to clean scale such as calcium carbonate, rust, and other debris from the wellbore and tubular. Hydrochloric acid mixtures are typically used for acid washing.

5.2.2 Fracking

Fracking, or hydraulic fracturing, is the process of blasting impermeable rock with a high-pressure fracking fluid to break the rock and create fractures. Fracking fluid is a mixture of roughly 90% water, 9.5% silica sand, and 0.5% additives. Millions of gallons of water are used per frack job. Water may be replaced by liquid nitrogen or liquid carbon dioxide. The sand acts as a proppant to keep the fractures open for hydrocarbons to flow through after pumping pressure is removed. It can be replaced with ceramic beads, pellets of aluminum oxide, or resin-coated sand.

The additives in fracking fluid have various functions. Friction reducers such as polyacrylamide reduce friction between the fluid and the pipe, while gelling agents thicken the fluid to keep sand particles in suspension. Other additives eliminate bacteria or prevent corrosion. Fracking fluid additives and their functions are summarized in Table 5.1. The specific makeup of fracking fluid varies depending on the reservoir. Concentrations of various additives may change, or some additives may be added or removed altogether.

Fracking can be performed on both open-hole and cased-hole perforations. It is carried out on reservoirs where permeability is very low such as tight gas reservoirs, shale formations, and some coal beds. Fracking can improve production by up to 30 times.

Improvements in fracking, combined with horizontal wells, led to the shale revolution in the United States where oil and gas in shale source rock could be produced in economic quantities for the first time. Figure 5.1 shows US gas production from various sources including shale. Shale production increased 60% from 2004 to 2020, and surpassed production from non-shale reservoirs.

The shale revolution enabled production of oil and gas from shale source rock which is impermeable and thin. The process of fracking creates fractures in the impermeable rock while the horizontal wells allow production of economic amounts of hydrocarbons. Perforations are made in the horizontal portion of the well and the frac fluid is pumped through the perforations into the rock.

Multistage fracturing is the creation of fractures in multiple stages along the wellbore in order to increase the surface area of the wellbore in contact with the reservoir. The fractures are spaced evenly apart.

Fracking design takes into consideration the type of rock, the depth of the reservoir, and the temperature at that depth. Computer models give the expected dimensions of the proposed fractures.

TABLE 5.1

Fracking fluid additives and their functions .

Additive	Function	Example(s)
Acid	Dissolves minerals and initiates fissures in the rock; removes drilling mud damage within the near-wellbore area	Muriatic acid, hydrochloric acid
Friction reducer (slickwater)	Reduces friction between pipe and fluid, allowing fracking fluid to be pumped to the target zone at higher rates and reduced pressures	Polyacrylamide, methanol, ethane-1,2-diol
Biocide	Prevents microorganism growth and reduce biofouling of the fractures	Glutaraldehyde carbonate, quaternary ammonium salts
Surfactant	Increases viscosity of fracking fluid to keep sand in suspension	Isopropanol
Gel	Increases viscosity of fracking fluid to keep sand in suspension	Hydroxyethyl cellulose, guar gum, methanol, ethane-1,2-diol
Crosslinker	Keeps fluid viscosity constant as temperature increases	Sodium borate, boric acid
Breaker	Reduces fluid viscosity by causing delayed breakdown of gel polymer chains	Ammonium persulfate
Stabilizer	Prevents corrosion of metal pipes	Oxygen scavenger, e.g., ammonium bisulfite
Gel stabilizer	Reduces thermal breakdown	Sodium thiosulfate
Scale inhibitor	Prevent scale building up in pipes	Ethylene glycol, inorganic phosphates
Corrosion inhibitor	Reduces corrosion in pipes	N,n-dimethyl formamide, propan-2-ol, methanol
Iron control	Prevents precipitation of metal oxides	Citric acid, ethanoic acid, acetic acid
Fluid loss additive	Prevents fluid loss and increases fluid efficiency	Diesel, sand
Buffer	Keeps pH of fracking fluid constant to maintain effectiveness of all additives	Sodium carbonate, ethanoic acid

The fracking process occurs as follows. First, frac fluid is mixed, then it is pumped into the well at high pressure to create cracks in the reservoir rock. The width of the cracks is on the order of millimeters. Fractures may extend tens or hundreds of meters vertically or horizontally, depending on the rock (Figure 5.2).

Fracking begins at the end of the lateral portion of the horizontal well, which is called the toe, and works back to the vertical portion of the well, one stage at a time. Horizontal wells are typically fracked in 10–15 stages depending on the lateral length. For vertical wells, fracking takes approximately one day. For horizontal wells, it can take 3–4 days for each well.

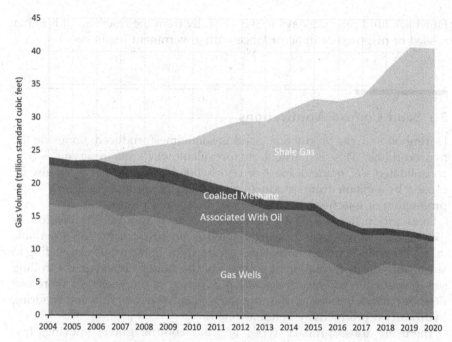

FIGURE 5.1
US annual natural gas production by source[1].

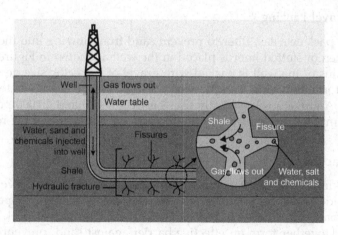

FIGURE 5.2
Hydraulic fracturing process.

After fracking is complete, fracking equipment is removed and replaced by production equipment. Hydrocarbons can now flow out of the reservoir into the well and to the surface. Along with the hydrocarbons, fluid from the frac job also flow out of the well. This is referred to as flowback fluid.

Flowback fluid can take days to recover fully from the reservoir. It is either reused or disposed of in accordance with government regulations.

5.3 Sand Control Applications

During oil and gas production, sand is sometimes produced alongside the hydrocarbons. This is particularly prevalent when the reservoir is unconsolidated, i.e., made of loose or poorly bonded sandstone. It can also be caused by erosion from production, or high pressures. Increased water production or injection also lead to increased sand production.

Sand production needs to be controlled as it can lead to erosion or plugging of the tubing and casing, as well as any instrumentation downhole or at the surface. Furthermore, it can create voids in the reservoir, leading to subsidence of the formation or collapse of the casing. However, controlling sand production can also lower the production of the well thus there is a delicate balance between protecting the well and the reservoir, and ensuring hydrocarbon productivity is not affected.

There are three primary types of sand control: gravel packing, frac packing, and chemical consolidation. They are described in more detail in the following sections.

5.3.1 Gravel Packing

A gravel pack acts as a filter to prevent sand from flowing into the well. A steel screen or slotted liner is placed in the well as shown in Figure 5.3 and the annulus of the well is packed with gravel of a specific size designed to prevent formation sand from flowing through. The screen holds the gravel in place. A gravel pack typically uses between 1,000 and 20,000 pounds of sand. Pumping pressures are limited to prevent fracturing of the reservoir.

5.3.2 Frac Packing

Frac packing is the simultaneous fracking and placement of a gravel pack to create wide and long fractures. The gravel pack prevents proppant from flowing back into the well and to the surface. The gravel pack and the proppant together form an effective barrier against sand, preventing sand production. Frac packs use 50,000 to 250,000 pounds of sand. Pump pressures are intentionally increased to exceed the reservoir fracture pressure.

5.3.3 Chemical Consolidation

Consolidation is used where other methods of sand control do not work well, for instance when there is a high-pressure differential between the

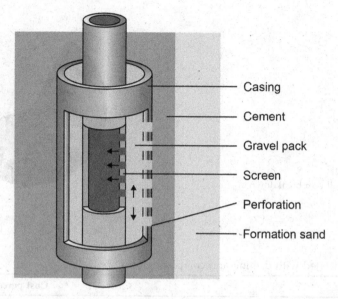

Casing
Cement
Gravel pack
Screen
Perforation
Formation sand

FIGURE 5.3
Gravel pack in a cased-hole well.

wellhead and the reservoir, or where there is high water production. Chemical consolidation arrests the sand in the reservoir as opposed to mechanical methods where sand is filtered at the wellbore. It involves mixing solid particulates such as quartz with a binding agent such as resin or cement, and a pore retaining agent like light oil, and injecting the mixture outside the casing into the sand-producing reservoir. This consolidates the unconsolidated sand preventing sand production. One limitation of chemical consolidation is that it can only be used for a single layer no thicker than 16 feet.

5.4 Completion Costs

Completion costs are typically broken down into materials and equipment for fracking, and sand control. Completion costs make up 55% to 70% of total well costs for onshore wells as shown in Figure 5.4. Typical costs for average onshore horizontal well are given in Table 5.2 with a combined drilling and completion cost of $7.5 million. Factors that drive the total cost include the location of the reservoir, depth of the reservoir, type of completion, design of the frac job, and lateral length. Cost percentages are similar for offshore wells.

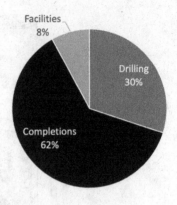

FIGURE 5.4
Average well cost breakdown[2].

TABLE 5.2

Costs associated with drilling and completing onshore wells

Category	Cost	Cost percentage (%)
Drilling		
Rig rates and drilling fluids	$1.28 million	17
Casing and cement	$0.98 million	13
Completions		
Hydraulic fracturing equipment	$1.95 million	26
Completion fluids and flowback disposal	$1.43 million	19
Proppants	$1.28 million	17
Facilities		
Separations equipment	$0.6 million	8
TOTAL	$7.52 million	

Drilling costs can range from $1.8 million to $2.6 million and cover rig rental, drilling fluids, casing, liner, and cement. Completion costs include water, additives, proppant, frac crews, wellhead equipment, completion tubing and liner, and rental of pumping equipment. Completion costs can range from $2.9 million to $5.6 million. Longer laterals require bigger frac jobs, leading to higher completion costs. Facilities costs make up 2% to 8% of total well costs, and are typically a few hundred thousand dollars. These costs include pumps or compressors to move hydrocarbons to gathering lines, separators, roads, evaporation pits, and flow lines.

After completions activities are over, the equipment is removed and production equipment installed to enable oil and/or gas to flow from the well. The well is now said to be producing and it becomes the responsibility of the production engineer.

Notes

1 *Natural Gas Gross Withdrawals and Production,* US Energy Information Administration, https://www.eia.gov/dnav/ng/ng_prod_sum_dc_nus_mmcf_a.htm
2 *Trends in U.S. Oil and Natural Gas Upstream Costs,* US Energy Information Administration, March 2016, https://www.eia.gov/analysis/studies/drilling/pdf/upstream.pdf

6

Production Engineering

Production optimization is the daily management of operations in order to maximize revenue from an oil field. Maximizing revenue requires minimizing operations costs and ensuring oil and gas production is uninterrupted. Marginal improvements to production volumes can have huge impacts since the life of a well can easily extend to 30+ years. Production engineers are the well doctors as they are responsible for repairing any well that goes down, i.e., loses production.

Production engineers carry out production optimization, well and reservoir monitoring, maintenance and repair, and plugging and decommissioning of wells at the end of their life.

6.1 Well Optimization and Reservoir Monitoring

Production data are used to evaluate well performance. Hydrocarbon production data from wells is often compared with production forecasts generated by reservoir engineers. Monitoring systems are set up so that if production falls below what is expected, production engineers can examine the well data to determine the root cause and how best to restore production.

Production engineers will confer with reservoir engineers before repairing a well. The reservoir engineer will determine how much oil and/or gas the well has remaining and will run economics to determine whether the remaining hydrocarbons can pay for the repair job and still produce a profit. Well tests can be conducted to help determine causes of reduced well performance. Sometimes fine-tuning or optimizing well equipment such as pumps, or debottleneck operations, can lead to incremental gains in production which in turn lead to significant profits over the lifetime of the well. Thus, production engineers are constantly optimizing production systems to maximize well production.

6.2 Wellbore Intervention

Wellbore interventions are repair or maintenance operations carried out to extend the life of a well or restore production. Over time, tubing may

DOI: 10.1201/9781003100461-6

corrode, deposits of scale or paraffins and other waxes may form, equipment downhole may fail, or the well may fill with water. All of these may negatively impact the performance of the well and may require intervention by production engineers. Often the production engineer will consult with the reservoir engineer to ensure that there are enough remaining hydrocarbons accessible to the well to pay for the cost of well interventions, and still produce a profit.

Interventions may be light or heavy. Light interventions are carried out without stopping production. They involve the use of slickline, coiled tubing, or wireline lowered into the well, which all minimize well blockage. A slickline is a single strand of wire, a wireline is a braided electrical cable, and coiled tubing is a long flexible metal pipe between 1 inch and 3.25 inches in diameter. Light interventions are useful for replacing or adjusting equipment downhole such as valves. They may be used to collect data downhole such as flow rates, temperature, and bottom-hole pressure.

Heavy interventions are also called workovers. They require a rig, called a workover rig, to remove the wellhead since production must be halted. Heavy interventions may be used to replace tubing or other equipment that cannot be retrieved using light interventions. Heavy interventions may also be used to plug and abandon producing zones so that the well can produce from secondary zones in what is known as a recompletion where a well is completed to produce from a new zone it did not previously produce from.

Artificial lift is used to lower pressure at the bottom of the well (bottom-hole pressure) in order to increase production rates of oil and/or gas. This can be done by placing a pump down at the bottom of the well. It can also be done through gas lift where gas from the surface is injected into tubing to reduce the density of fluids in the tubing from the reservoir.

Another form of artificial lift is plunger lift which uses a free piston to travel up and down in the tubing of the well. The piston creates a mechanical seal between the fluids above and below the plunger. Its movement up and down the well not only helps to control gas production but also helps to scrape initial deposits of scale and paraffin, lifting them to the surface. Production from gas wells can be lowered by removing liquids from the reservoir. Artificial lift removes liquids from the well and the reservoir so that gas can flow at higher rates.

6.3 Offshore Production Platforms

The first offshore rigs were built in the 1940s in less than 30 feet of water on pipes connected to the land. Today, offshore rigs can extract oil and gas from depths of up to 10,000 feet. Floating platforms are used in water deeper than 1,300 feet as construction of fixed rigs at those depths is expensive.

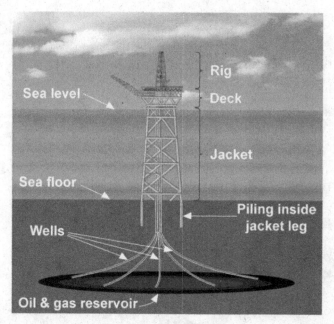

FIGURE 6.1
Parts of an offshore platform[1].

The different parts of an offshore production platform are shown in Figure 6.1. The deck and the rig containing production equipment are found above sea level. The deck is a multilevel structure that contains equipment and living quarters. Under the deck is a jacket which is a tall vertical section made of tubular steel. The jacket provides stability and support for the entire structure, as well as protection for the internal piping and equipment. Steel pipes called piling are driven through the legs of the jacket to secure the platform hundreds of feet into the sea floor. Wells are drilled into the sea floor. Directional drilling is used to drill and produce multiple wells from a single platform.

Different types of offshore production platforms are shown in Figure 6.2. Floating offshore production systems are permanent but floating production structures in deep water. One type of floating system is the Floating, Production, Storage, and Offloading (FPSO) system which is used for both production and storage at sea. It comprises a large ship moored with a rope with periodic offloading of hydrocarbons through shuttle tankers that carry the hydrocarbons to shore.

Tension leg platforms (TLP) are used for depths of up to 5,000 feet. The platform consists of pontoon-like steel pillars filled with air that are adjusted via connected cables to provide weightlessness to the structure above the water.

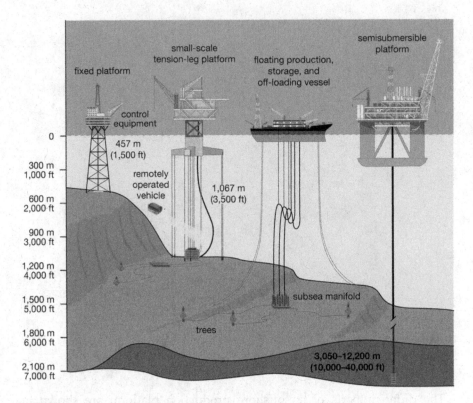

FIGURE 6.2
Offshore production platforms[2].

Spar platforms can be used in depths up to 7,500 feet. They are moored to the seabed and weighted at the bottom to keep them oriented upright. A spar platform can be moved horizontally and positioned over wells away from the main platform location.

Semisubmersible platforms can operate at depths up to 5,000 feet. Some are transported using tugs and barges, while others have self-propulsion. They have columns that are submerged, making the rig more stable, and lessening rolling and pitching.

In addition to large platforms, subsea platforms can be used for offshore hydrocarbon production. Subsea completions systems are steel frames with water-tight equipment and processing facilities mounted to them. The component pieces of equipment include compressors, pumps, and separators. They are placed on the sea floor and connected to each other to form a large production unit. Subsea completions systems are deployed in deep or ultra-deep parts of the sea.

Because subsea equipment is closer to the source than in other offshore platforms, separation and processing of hydrocarbons is simpler and more efficient. The processed fluids are then sent onshore or to an offshore platform. In this way, multiple wells may be connected to a single platform, thus making production cheaper and more efficient.

6.4 Well Plugging and Abandonment

Depending on the type of reservoir, a well can keep producing economic quantities of oil and/or gas for 15 to 30 years on average. At the end of life of a well, it must be plugged and abandoned (P&A) to prevent fluids flowing to the water table or to the surface. Dry holes which do not produce hydrocarbons must also be properly plugged and abandoned. The well casing is filled with cement, and all equipment at the wellhead is removed. The wellbore is then plugged with cement at multiple depths to prevent fluids moving from the reservoir into the wellbore. The specific requirements for plugging and abandoning a well depend on the laws of the state or country that the well is in.

Bioremediation of the wellsite is carried out to restore the environment as much as possible to its original condition. Microorganisms are used to digest hydrocarbon contamination. In Pennsylvania, sediment and drill cuttings are sometimes mixed with cement to produce bricks.

Decommissioning offshore wells entails plugging all the wells supported by the platform and severing their well casings. The platform is dismantled and taken to land so it can be refurbished and reused. For platforms in deeper water, say over 100 feet, all useful equipment is removed for later use. The platform is toppled at the site or at a new location and allowed to sink to the ocean floor. It becomes a habitat for various types of marine life including coral, sponges, and grouper.

6.5 Operations Costs

The costs associated with the production phase are fixed lease costs, variable operating costs, and plugging and abandoning (P&A) or decommissioning costs. Well type (oil and/or gas), location, performance or amount of production, and life of the well determine total operating expenses. The fixed lease costs, also called lease operating expenses (LOE), include costs associated with maintenance, artificial lift, and other wellbore intervention activities. LOE is incurred over the life of the well and depends on the reservoir. Costs increase as hydrocarbon production and depth of well increase. LOE is usually reported on a $/BOE basis.

For onshore wells, typical lease operating costs over 20 years run from $1 million to $3.5 million. LOE costs for onshore wells range from $2/BOE

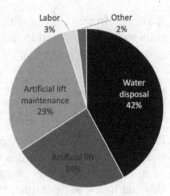

FIGURE 6.3
Typical distribution of LOE costs in the Bakken shale play.

to $14.50/BOE with water disposal costs included. Water disposal costs can range from $1/bbl to $8/bbl of water. In the Bakken shale play, costs related to artificial lift typically make up over 50% of LOE costs, while pumping and compression costs make up 2% of total LOE. Bakken LOE costs are shown in Figure 6.3.

Variable operating costs are related to delivery of oil and gas including gathering, processing, and compression. As these costs vary depending on the volume being delivered, they are usually measured in $/MCF or $/bbl. They vary depending on the type of hydrocarbon. These costs are covered in the Facilities Engineering chapter.

Typical P&A costs for onshore wells are around $1 million. This amount can vary depending on the well type (vertical or horizontal) and well depth, among other factors.

Offshore operating costs are variable costs and are generally tied to platform operation and maintenance, and delivery of oil and gas to purchase points. Costs vary depending on various factors including water depth, facility size, and distance from the shore. The semisubmersible platform is among the most expensive to operate at $9 billion over a lifetime of roughly 80 years, with a fixed lease cost of $17 per barrel of oil equivalent (BOE). Operating costs for the TLP are much lower at $2.1 billion over the same lifetime, and a fixed lease cost of $10/BOE. Lifetime offshore platform costs are summarized in Table 6.1.

TABLE 6.1

Total operating costs for offshore platforms assuming a lifetime of roughly 80 years[3]

Offshore platform	Lifetime operating costs (Million $)	Lifetime operating costs ($/BOE)
Semisubmersible	9,000	17
Spar	2,500	9
TLP	2,100	10
Subsea	500	8

Decommissioning costs for a semisubmersible with recycling of equipment can be on the order of $30 million. For a spar platform toppled at the site to form a marine habitat, decommissioning can cost about $15 million.

Notes

1 *Geologic Energy Management – Offshore Platform*, CA Department of Conservation, https://www.conservation.ca.gov/calgem/picture_a_well/Pages/offshore_platform.aspx
2 Offshore Drilling Platforms, *Encyclopaedia Britannica*, https://www.britannica.com/technology/petroleum-production/Deep-and-ultradeep-water#/media/1/1357080/252533
3 *Trends in U.S. Oil and Natural Gas Upstream Costs*, US Energy Information Administration, March 2016, https://www.eia.gov/analysis/studies/drilling/pdf/upstream.pdf

7

Facilities Engineering

Facilities engineers are responsible for handling design and construction of processes, facilities, and equipment to measure, control and purify hydrocarbons. Purification of oil and gas is carried out to remove impurities such as sand, carbon dioxide, and water. It is done to get oil and gas to specifications requested by downstream buyers and to prevent corrosion of pipelines during transportation of hydrocarbons. Facilities engineers also handle operation, maintenance, and repair of these facilities and equipment.

Facilities engineering can overlap with midstream oil and gas which involves transportation of hydrocarbons from the separation facilities to customers downstream.

7.1 What is Facilities Engineering?

Upon completion of a well, the facilities engineers design facilities at the wellhead and at central gathering locations to separate hydrocarbons from unwanted liquids, gases, and other materials that also come up from the subsurface. The gathering systems consist of pipelines from multiple wells to central points which may be larger pipelines, storage facilities, compression stations, processing plans, or shipping-off points. Compression increases pressure thus maximizing gas production from wells, and providing the driving force for gas along pipelines.

Separation is necessary because oil is produced with water, sand, and may also be produced with gases such as natural gas, carbon dioxide, and hydrogen sulfide. Natural gas may also need to be separated from unwanted gases such as carbon dioxide, hydrogen sulfide, and water vapor.

Facilities are processing plants that are needed to separate oil and gas from impurities so that they can reach the specifications of the customers downstream. Transportation of hydrocarbons is usually done via pipeline. Impurities such as carbon dioxide and hydrogen sulfide gas can mix with water to form acids which then corrode the pipelines, compromising their integrity. Thus, facilities engineering is a critical part of oil and gas engineering.

Purification of hydrocarbons is carried out in stages. First, the solid components such as sand are removed. Then, the liquids are separated from

DOI: 10.1201/9781003100461-7

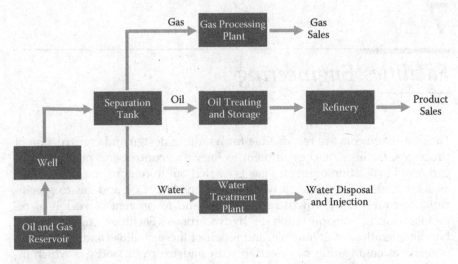

FIGURE 7.1
Oil and gas production from reservoir to sales.

the gases. Separations at the wellhead may be powered by the combustion of gaseous hydrocarbons, or through the use of solar panels.

Facilities are needed for both onshore and offshore operations. The type and concentration of impurities determine the facilities that are needed.

After crude oil, gas, and water are separated from each other, the gas is sent to a gas processing plant where the different gases are separated from each other and from gaseous impurities. Crude oil is sent to a refinery where it is separated into different products which are then sold. Water is treated and disposed of or injected into water injection wells. This process is summarized in Figure 7.1.

7.2 Midstream

As the name implies, midstream oil and gas falls between upstream and downstream. It involves the storage, marketing, and transportation of oil and gas between upstream and downstream. This includes pipelines and tankers. Other key services offered by midstream include blending of petroleum products and injection of additives

The network of gas transmission and hazardous liquids pipelines in the US in 2021 is shown in Figure 7.2. Hazardous liquids in this context refers to crude oil, condensate, and products of oil refining.

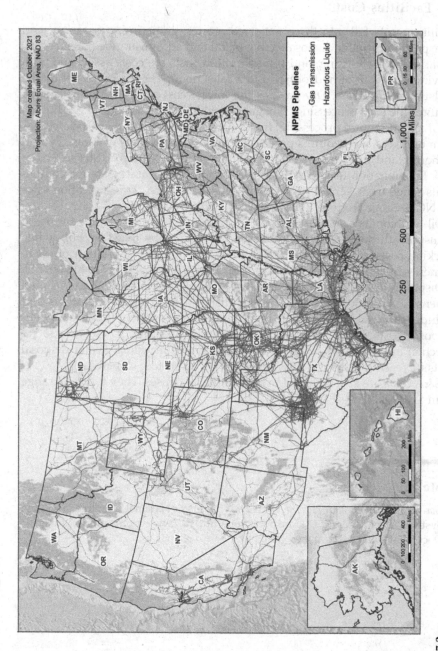

FIGURE 7.2
Pipelines across the United States in 2021[1].

7.3 Facilities Costs

Facilities costs fall into three major categories: gathering, processing, and transportation; water disposal; and general and administrative (G&A) costs.

Gathering, processing, and transportation costs are costs associated with moving each unit of hydrocarbon to a sales point. These fees are set by the midstream providers. Dry gas has the lowest cost at $0.35/Mcf because it requires the least processing. Wet gas is gas that contains natural gas liquids (NGLs) that require processing, fractionation, and transport. Associated gas from oil plays also requires processing and is classified as wet gas. Typical gathering and processing fees for wet gas range from $0.65/Mcf to $1.30/Mcf. Fractionation fees to separate NGLs into its separate gaseous components range from $2/bbl to $4/bbl of NGL recovered. Transportation rates for NGLs range from $2.2/bbl to $9.78/bbl of NGLs.

Oil and condensate are transported through gathering lines or by truck. Transport through gathering lines costs $0.25/bbl to $1.50/bbl, while trucking costs range from $2.00/bbl to $3.50/bbl. Transportation of oil to refineries by pipeline or rail has a price differential of $2.20/bbl to $13.00/bbl.

Disposal of flowback water from well drilling initially falls under capital expenses. After 30 to 45 days, water disposal falls under operations and includes disposal of water from the formation and residual flowback. Water disposal costs range from $1/bbl to $8/bbl of water. They include re-injection into water disposal wells, trucking, and recycling. Water disposal costs increase as the ratio of water to oil or water to gas increases.

G&A costs are also included in operational expenses. They typically range from $1/BOE to $4/BOE.

Note

1 *Gas Transmission and Hazardous Liquid Pipelines*, National Pipeline Mapping System, Department of Transportation, https://www.npms.phmsa.dot.gov/Documents/NPMS_Pipelines_Map.pdf

8

Oil Refining

Refining is the downstream part of the oil and gas industry. Crude oil from upstream operations is sent via midstream transportation infrastructure such as pipelines to refineries downstream. Refining is carried out to get useful products out of crude oil for use as fuel and feedstock for production of various chemicals. As described in Chapter 1, products from the refining of petroleum are used for materials, heat, electricity, fuel, and petrochemicals, among others.

8.1 Gas Processing

Gas processing is the gas equivalent of oil refining. Gas mixtures comprise methane and natural gas liquids (NGLs). These mixtures come from gas wells or oil wells. Gas found in oil wells is referred to as associated gas. It can either exist separately from crude oil, or it can be dissolved in it. Besides hydrocarbon gases, gas mixtures can contain impurities such as water vapor, hydrogen sulfide, and carbon dioxide. The raw mixture is sent from upstream to a gas fractionation plant where it is heated and cooled to separate the components into individual gases such as ethane, propane, butane, isobutane, and heavier gases. This process is summarized in Figure 8.1.

Hydrogen sulfide and carbon dioxide are removed using amine liquids in a process called sweetening. This process is further described in Section 8.4.

Methane, or natural gas, is separated from NGLs using cryogenic expansion whereby the temperature of the mixture is rapidly lowered to −120°F. The temperature drop condenses the NGLs while leaving methane as a gas. The gases in the NGL mixture are separated from each other using fractionation where the components are boiled off one at a time.

Natural gas is odorless. Methyl mercaptan, an odorant that smells like rotten eggs, is added to natural gas to provide a warning if leaks happen.

8.2 Refining Processes

Refining processes fall into three basic categories: separation, conversion, and treating. Crude oil components are separated from each other using heat.

DOI: 10.1201/9781003100461-8

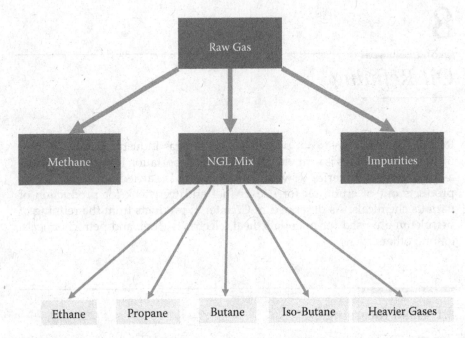

FIGURE 8.1
Natural gas processing.

They are then converted to more valuable products and treated to remove impurities such as sulfur.

8.2.1 Separation

Separation, or distillation, is done by heating crude oil in a distillation tower at temperatures up to 800°F at the bottom of the tower. Temperatures at the top of the tower may be as low as 300°F. Distillation is done at atmospheric pressure for lower temperatures, and under vacuum for the highest temperatures. Crude oil is a mixture of hydrocarbons of different chain lengths, which leads to them having different boiling points. The longer the chain, the higher the boiling point. At different temperatures, different crude components, also called fractions, boil off and are collected, as illustrated in Figure 8.2.

Methane, ethane, and propane are either collected for use as fuel gas in the refinery or are flared. Propane and butane, together called liquified petroleum gas (LPG), are collected at the top of the tower, pressurized, and sold. Light fractions such as naphtha boil off first at the top of the tower. Medium-weight liquids such as kerosene separate in the middle part of the tower, and heavier fractions such as asphalt settle at the bottom of the tower.

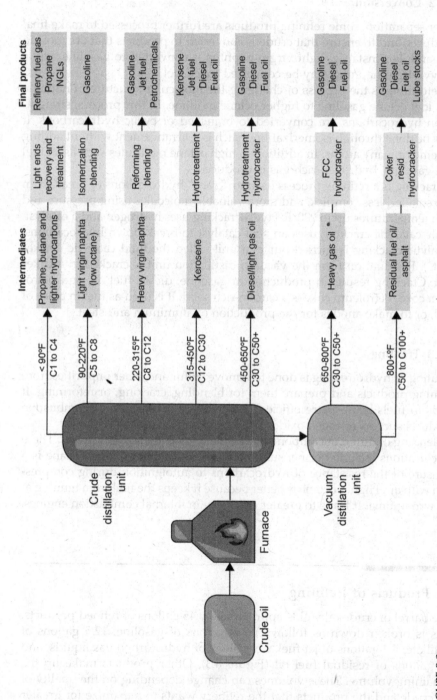

FIGURE 8.2
Products of refining of crude oil.

8.2.2 Conversion

After separation, some refining products are further processed to make final products, and to ensure that crude is converted to products that are most in demand. For instance, light virgin naphtha is converted to gasoline, while heavy virgin naphtha may be converted to jet fuel.

Reforming is the process of changing the molecular structure of naphtha and low octane gasoline to higher octane gasoline. In this process, straight-chain hydrocarbons are converted to branched or cyclic hydrocarbons. It may be done through isomerization which is rearrangement without adding or removing any atoms. In addition to high octane molecules such as jet fuel and gasoline, hydrogen-rich gas is produced.

Cracking is a refining process used to "crack" hydrocarbons or break them into smaller, less complex, and more valuable molecules using pressure and high temperatures up to 930°F. Hydrocracking uses hydrogen and a catalyst, while catalytic cracking uses an acid catalyst to break down hydrocarbons. Catalytic cracking is carried out in a unit called the fluid catalytic cracker (FCC). Residual oil from the vacuum distillation unit is cracked in a coker unit. Cracking results in production of gasoline, diesel, fuel oil, and petroleum coke. Petroleum coke is a carbon-rich solid. It is used as fuel in place of coal, or to make anodes for the production of aluminum and steel.

8.2.3 Treating

Treating or hydrotreating is done to remove sulfur and other impurities from refining products and prepare them for blending, cracking, or reforming. It leads to fuels being more efficient and reduces production of combustion products such as nitrogen oxides (NOx) and sulfur oxides (SOx).

Blending of different components is done to meet fuel specifications. These specifications include octane, sulfur content, and boiling point. Octane is a measure of the resistance of hydrocarbons to autoignition during compression with air. Higher octane is better because it keeps the fuel from igniting at the wrong time. It leads to greater efficiency in internal combustion engines.

8.3 Products of Refining

One barrel of crude oil yields approximately 45 gallons of refined products. This is broken down as follows: 19.4 gallons of gasoline, 12.5 gallons of distillate, 4.4 gallons of jet fuel, 1.5 gallons of hydrocarbon gas liquids, and 0.5 gallons of residual fuel oil (Figure 8.3). Other products make up the remaining volume. These volumes can change depending on the quality of the crude and the products that the refinery wants to maximize for greater

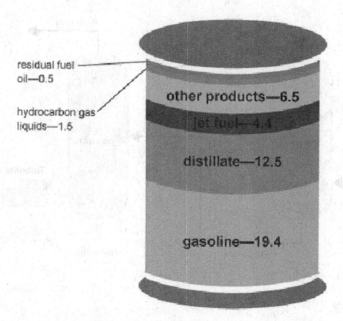

FIGURE 8.3
Petroleum products from a typical barrel of crude oil[1].

profit. The products listed can be converted to other products as has been described in the sections above.

Liquified petroleum gas (LPG) is used for heating and cooking in the home. It may also be used to power vehicles and generators. Diesel and gasoline are used for powering vehicles. Naphtha is used as feedstock for production of petrochemicals. Heavy gas oil is used mainly for heating or powering engines. Kerosene is used as jet fuel. Asphalt is used for paving roads while residual fuel oil is used to power ships.

8.4 Pollution from Refineries

Leaks from refinery processes and storage tanks can result in the release of chemicals into the air, water, or soil through equipment and line leaks.

Carbon dioxide (CO_2) typically accounts for 98% of greenhouse gases emitted by refineries, followed by methane (CH_4) with 2.25%, and dinitrogen oxide (N_2O) with 0.08%.[2] These amounts may vary depending on the specific processes occurring at a refinery. Leaks may be mitigated through leak detection using sensing technologies, and repair and replacement of lines and equipment. Light Detection and Ranging (LiDAR) is one example of a leak detection technology. It uses eye-safe lasers to

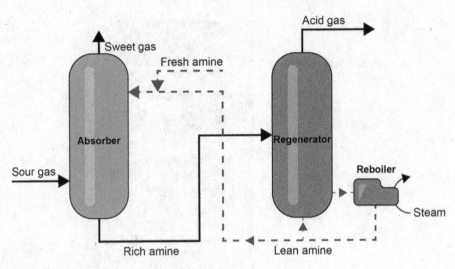

FIGURE 8.4
Schematic of an amine unit.

create topographic or gas concentration imagery. Thus, it can detect leaks, as well as quantify the leak rate. LiDAR and other sensor technologies like infrared cameras may be combined with drones to reduce manpower needed to detect leaks.

Sour gas, or gas containing high amounts of sulfur compounds, is a by-product of refining. It comprises sulfur dioxide (SO_2), carbon dioxide (CO_2), hydrogen sulfide (H_2S), and may contain some hydrocarbons. Sour gas is sent to an amine treatment unit where SO_2, CO_2, and H_2S are absorbed by the amine, producing a sweet gas, i.e., low sulfur gas.

This process is illustrated in Figure 8.4. Sour gas enters an absorber which contains a liquid called an amine. The amine absorbs the acid gases which are CO_2, H_2S, and SO_2, and is then called rich amine. The rich amine is sent to a regenerator unit where it is heated up, resulting in the acid gas being released from it and collected. The amine is then cooled and sent back to the absorber, and the cycle repeats.

Other pollutants that may be released to the air from refineries are benzene, toluene, ethylbenzene, and xylene, which are collectively called BTEX compounds. Refineries are also sources of criteria air pollutants such as particulate matter (PM), carbon monoxide (CO), and nitrogen oxides (NO_x). NO_x combines with volatile hydrocarbons to form ozone.

Wastewater from refineries may contain hydrocarbon residue and other hazardous waste. It is sometimes disposed of in deep injection wells. Ground and water pollution may come from spills and equipment leaks, sludge from refinery processes, and used catalysts.

Notes

1 *Oil and Petroleum Products Explained*, US Energy Information Agency, https://www.eia.gov/energyexplained/oil-and-petroleum-products/
2 *Available and Emerging Technologies for Reducing Greenhouse Gas Emissions from the Petroleum Refining Industry*, US Environmental Protection Agency, 2010, https://www.epa.gov/sites/default/files/2015-12/documents/refineries.pdf

Glossary

Term	Definition
Aquifer	An underground rock layer saturated with water.
Artificial lift	Techniques used to lower pressure at the bottom of the well in order to increase production rates of oil and gas.
Condensate	Hydrocarbon liquids that condense out of the gas stream into a liquid that can be stored at room temperature.
Crude oil	A liquid mixture of hydrocarbons formed underground from the remains of plants and marine life such as algae and plankton.
Drive mechanisms	The natural energy of the reservoir that can be used to drive hydrocarbons into the wellbore.
Flaring	Burning gaseous hydrocarbons under controlled conditions.
Fracking	The process of blasting impermeable rock with a high-pressure fracking fluid to break the rock and create fractures.
Horizontal drilling	A type of drilling that starts with a vertical well that turns horizontal as it nears the reservoir.
Hydrocarbon	A naturally occurring compound consisting of carbon and hydrogen.
Hydrocarbon saturation	The percentage of fluids in the reservoir that are hydrocarbons.
NGLs	A combination of condensate and condensed gaseous liquids captured at natural gas plants.
Offshore drilling	Drilling to reach reservoirs below the sea bed.

OHIP	The quantity of hydrocarbons initially in the reservoir before the production begins.
Permeability	Connected pores which provide pathways for fluid flow.
Petroleum	A liquid mixture of hydrocarbons formed underground from the remains of plants and marine life such as algae and plankton.
Porosity	The percentage of void space in a rock which can hold liquids or gases.
Recovery factor	The percentage of original hydrocarbons in place that can be economically produced.
Refining	The process of breaking down and transforming crude oil into useful products.
Reservoir	A body of rock with sufficient porosity and permeability to transmit fluids.
Reservoir engineering	The engineering discipline that deals with the transfer of fluids to, from, or through a reservoir.
Shale	A laminated sedimentary rock that comprises fine silt, clay, and other minerals.
Sour gas	Gas containing high amounts of sulfur compounds.
Sweet crude	Oil that contains less than 1% sulfur.
Triple combo log	A combination of gamma ray, resistivity, and density-neutron logs.
Unconventional reservoir	A reservoir that requires special recovery operations outside conventional operating practices.
Well log	A detailed record of changing properties with depth in a wellbore.

Index

Printed in the United States
by Baker & Taylor Publisher Services